CPU Design:
Answers to Frequently
Asked Questions

This page intentionally left blank

Chandra M. R. Thimmannagari

CPU Design: Answers to Frequently Asked Questions

Springer

eBook ISBN: 0-387-23800-X

Visit Springer's eBookstore at: http://ebooks.springerlink.com
and the Springer Global Website Online at: http://www.springeronline.com

To my wife Haritha,
my kids Sanjana and Siddharth,
and my parents Krishna Reddy and Saraswathi

This page intentionally left blank

Contents

This page intentionally left blank

Foreword

I am honored to write the foreword for Chandra Thimmannagari's book on CPU design. Chandra's book provides a practical overview of Microprocessor and high end ASIC design as practiced today. It is a valuable addition to the literature on CPU design, and is made possible by Chandra's unique combination of extensive hands-on CPU design experience at companies such as AMD and Sun Microsystems and a passion for writing.

Technical books related to CPU design are almost always written by researchers in academia or industry and tend to pick one area, CPU architecture/Bus architecture/ CMOS design that is the area of expertise of the author, and present that in great detail. Such books are of great value to students and practitioners in that area. However, engineers working on CPU design need to develop an understanding of areas outside their own to be effective. CPU design is a multi dimensional problem and one dimensional optimization is often counterproductive.

For instance, as someone who mainly does CPU architecture, I have found that CPU architects who understand how logic design, circuit design and chip integration are really done in practice do a much better job architecting the chip. There are constraints in these different areas that could make an architectural idea hard to implement, and an architect who understands these constraints is more likely to make the right decisions upfront. However, there are really no books out there to help an architect understand quickly how the later stages of chip design work. Reading detailed technical books on physical design to obtain this knowledge is typically not an option given time constraints. The most accessible way today to learn the broader skill set necessary is from chatting with friends and picking up bits of knowledge here and there. Over time the good ones do develop a working knowledge of all areas of CPU design, but it takes many years.

The same goes for circuit designers who want to understand architecture. I have had several circuit engineers come to me wanting to know more about architecture. I answer questions as time permits and suggest they read "Hennessy and Patterson". It helps, but goes only so far even though H&P is a really well written book. There is just too much detail, and it is hard to filter out what is relevant.

In a way Chandra's book is structured as a chat with a knowledgeable friend with much time to spare. So we could imagine a circuit designer who is working on a cache, and has a design problem - for instance, the replacement algorithm he is trying to implement is not making timing. He will have to discuss this with the logic owner or architect, but it will help if he has an understanding of the architectural options available and any potential circuit issues with those options prior to the discussion. He could look up this book and starting with the first question on caches (Q5 in Architecture: *What is cache memory in a CPU and what are the most common terms associated with caches?*) work through replacement policy related questions (Q10 to Q15 in Architecture) to develop an understanding of the options available. Or imagine an architect who is told that the particular idea she has in mind cannot be implemented owing to routing density issues related to noise. She could look up the relevant question in the book (Q6 in Circuits and Layout: *What do you mean by effect of noise in a design and what are the most common techniques used to reduce its effect?*) to develop a quick understanding of noise issues as well as possible solutions and work with designers to find a way to implement her idea.

The book also provides excellent lists of techniques in the experienced logic/circuit designer's toolbox to attack a problem. For instance, a logic designer who is trying to figure out how to make timing for a block could go straight to Q4 in the Logic chapter and look at the list of suggestions there for fixing timing paths and start making headway. Or a designer who is trying to reduce power for a block or a chip could go to Q6 in the Logic chapter and look at the list of suggestions there for reducing power. Or a circuit designer who is trying to fix noise problems could go to Q6 in the Circuits and Layout chapter. Or a manager who wants to learn about design tools available for a particular task could go to the relevant question in the Tools chapter.

The book also includes good, concise descriptions of many thorny issues in CPU design such as RAS, electromigration, IR drop, pass gate muxes and mintime fixes.

I believe the book will be a valuable addition to any CPU designer's library.

Rabin Sugumar
CPU Architect
22 September 2004

Preface

This book describes the basic concepts and techniques used towards building a Microprocessor. This book is made primarily for graduate students and design engineers as a quick reference material. Readers will be exposed to an effective processor design methodology. Some of the things covered here are, techniques to fix timing of a critical path, techniques to reduce power dissipation in a block, typical processor design flowchart, concepts of caches, techniques to fix mintime violations, techniques to fix noise violations, concepts of flops and latches, various multithreading techniques used in a processor design, various benchmarks used for CPU performance evaluation, various tools used in a processor design, concepts of Verilog, some Verilog coding guidelines, implementation details of out-of-order processor, concepts of electromigration and IR drop etc.

CHANDRA M. R. THIMMANNAGARI
Senior Staff Engineer

This page intentionally left blank

Acknowledgements

I would like to express my heartfelt gratitude to a number of people who contributed their time and effort towards this book. Without their help, it would have been impossible to take this enormous task.

First and foremost, I would like to thank my family, who gave me continuous support and encouragement that kept me constantly motivated towards the completion of this project. A special thanks to my wife Haritha, who patiently withstood my late night book activities. I could not have accomplished this task without her help and understanding. Thank you Haritha.

I am extremely fortunate to have it reviewed by Rabin Sugumar whom I have always respected for his extraordinary technical skills and down to earth personality. Thank you Rabin.

I would like to thank the following people who have devoted their precious time in reviewing the manuscript.

Rambabu Pyapali (Sun Microsystems)

Mathew Joseph (Sun Microsystems)

Suresh Tirumalaiswamy (Sun Microsystems)

Bruce Petrick (Sun Microsystems)

Ramaswamy Sivaramakrishnan (Sun Microsystems)

Sunil Vemula (Sun Microsystems)

Bob Nuckolls (Sun Microsystems)

Grant Davidson (Sun Microsystems)

Sorin Iacobovici (Sun Microsystems)

Amjad Qureshi (Cradle Technologies)

Andy Charnas (Cradle Technologies)

George Phan (Cradle Technologies)

Jason Lin (Cradle Technologies)

Sudhakar Bhat (Intel Corporation)

Ravisekhar Reddy Naral (Texas Instruments)

Jagan Mohan Reddy Thimmannagari

I would like to thank all my previous employers, Advanced Micro Devices, Sun Microsystems and Cradle Technologies for giving me an opportunity to work on the latest cutting edge technologies.

Also I would like to thank Michael Hackett, Rebecca Olson and Deborah from Kluwer Publishers in supporting me throughout the project.

1 Architecture

1. What are some of the responsibilities of a Chip Architect?

Some of the responsibilities of a Chip Architect are summarized below.

Table 1: Architect Responsibilities

1	Do market research to see what kind of Chip he wants to build (i.e whether he should be building a chip targeting commercial applications (database etc.) or technical applications (applications which need high performance computing such as DSP applications, BioTech applications etc.) or both commercial as well as technical applications) or talk to his/company's customers to see what they are interested in.
2	Come up with the feature set for the Chip i.e ask the following questions to himself (below questions mostly relate to Chip being a Microprocessor) 1. Should it be a CMP(Chip Multiprocessing) or a Non-CMP Chip. 2. Should the Chip be supporting more than one Thread i.e should it be Single Threaded or Multi Threaded. 3. If it is Multi Threaded how many Threads it should be supporting and should it be supporting SMT (Simultaneous Multithreading) or VT (Vertical Threading). 4. Should the Core part of the Chip be Out-Of-Order or In-Order. 5. Should it be Superscalar or Non-Superscalar and if it is Superscalar how wide it should be. 6. What should be the Frequency target i.e the frequency at which it should be operating at. 7. Number of Pipe Stages it should be supporting. 8. Should it be supporting Static/Dynamic Branch Prediction. 9. Number of Cache Levels, Size of various Caches, Cache Policies and Associativity. 10. Size of Main Memory it should be supporting. 11. Should it be designed for SMP (Symmetric Multiprocessing) systems or Uniprocessor Systems. 12. Should it provide support for an on-chip Memory Controller and if it does provide support for an on-chip Memory Controller how many Memory Controllers it should be supporting. 13. What should be the Peak Memory bandwidth it should be supporting. 14. What should be the Power Spec for the Chip (i.e Peak Power number and Average Power number). 15. Amount of RAS (Reliability, Availability and Serviceability) it should be supporting. 16. What should be the external interface to the outside world (i.e the kind of System Bus it should be supporting etc.).
3	Come up with a high level (i.e in C/C++) Performance Model for the Chip and run various benchmarks to see if the performance numbers meet the Performance Spec defined for the Chip.

Table 1: Architect Responsibilities

4	Partition the Chip into several major Blocks and come up with a 1 pager document for each of the blocks describing its functionality and the number of pipe stages within it. Pass on these documents to the Microarchitects (RTL/Logic designers).
5	Get constant feedback from the Microarchitects regarding information related to critical paths within there blocks and try to see if anything could be done at the architectural level to resolve them without compromising much on performance i.e adding few pipe stages, reducing the operating frequency etc. Feed the proposed changes into the performance model to see if we can live with the performance impact of these changes.
6	Get constant feedback from the Circuit designers and Integration folks regarding information related to Power dissipated and Area occupied by the Chip. If these numbers are beyond the numbers provided to them then try to see if anything could be done at the architectural level to resolve them without compromising much on performance i.e reducing the Issue width, reducing the number of Thread support, reducing the Cache size, lowering the Frequency etc. Feed the proposed changes into the performance model to see if we can live with the performance impact of these changes.
7	Present the Spec to various groups within the Company i.e System groups, Design groups etc.
8	Work with the System folks to resolve any System issues with the Chip.
9	Come up with the Programmer's Reference Manual.
10	Present the Spec at various well known Conferences mainly trying to gain support/momentum for the Chip.
11	Patent all the novel Architectural ideas which you were part off and which got implemented in the Chip.

2. Describe a typical ASIC design flow using a Flowchart?

A typical ASIC design flow as shown in Figure 1 has the following steps -

1	Spec out the Architectural definition of the ASIC.
2	Code the Functionality described in the Spec using hardware descriptive language (i.e Verilog or VHDL).
3	Compile the Code and Verify its functionality by simulating the Code (i.e RTL model) using one of the HDL simulators in a test bench environment.
4	Synthesize the Code by applying proper Area and Timing constraints and generate a Gate Level netlist.
5	Do equivalence checking between the Gate Level netlist and the RTL model.
6	Floorplan the design using one of the in-house or commercially available floorplan tools.
7	Place and Route all instances within the design using one of the in-house or commercially available Place and Route tools.

8	Run Physical verification by running LVS (Layout versus Schematic), DRC (Design Rule Checking) and ERC (Electrical Rule Checking) on the Layout database.
9	Extract Parasitics.
10	Extract Gate level netlist and verify its functionality by running Gate Level simulation using one of the HDL simulators.
11	Do equivalence checking between the extracted Gate level netlist and the coded behavioral RTL.
12	Run Static Timing Analysis to find the Maxtime and Mintime paths. If there are any such paths then loop back to RTL (if the paths cannot be fixed by Re-Synthesis) or Logic Synthesis (if the paths cannot be fixed by Re-Floorplaning and Redoing Place & Route) or Floorplanning (if the paths can be fixed simply by Re-Floorplaning and Redoing Place & Route) and repeat the steps in the flowchart all over again.
13	Tapeout the Chip once the Functionality is verified and Timing spec is met.

Figure below shows the flowchart for a typical ASIC design flow.

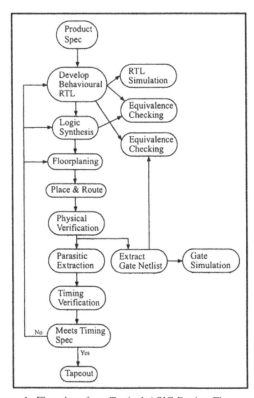

Figure 1: Flowchart for a Typical ASIC Design Flow

3. Describe a typical Processor Design flow using a Flowchart?

A typical Processor design flow as shown in Figure 2 has the following steps -

1	Define the Processor Spec (i.e Performance numbers (i.e SpecInt, SpecFp, TPC-C etc.), Die Size, Power Consumption, Process Technology, Frequency, Pin Count, Type of Package, Uniprocessor/MultiProcessor Support etc.)
2	Spec out the Architectural definition of the Processor (i.e number of Pipe Stages, number of Cache Levels, type of Cache Coherency Protocol, Cache Size, Line Size, Size of Main Memory, Page Size, Out-Of-Order/In-Order, Superscalar/Non-Superscalar, Support for Multiple Threads etc.)
3	Develop a high level Performance Model and run various benchmarks to see if the results meet the desired performance goals.
4	Partition the Processor into several major Blocks and each major Block into several Sub-Blocks (i.e Control, Datapath and Megacells).
5	Define Microarchitectural Spec for each of the major Blocks (i.e number of Pipe Stages within the Block, Functionality within each stage, number of Sub-Blocks, Area of the Block etc.)
6	Analyze the paths within the Block to see if there are any paths which will be timing critical. Come up with a list of such paths and simulate them using dynamic timing analysis tools (i.e Spice) to see if they meet timing. If the paths fail to meet timing with all the design optimization tricks (i.e using dynamic gates, shielding signal lines etc.) then talk to the Architect to see if he can update the Spec (i.e repartitioning some of the functionality between the Blocks, Simplifying few architectures features etc.) to take care of such paths. This is an iterative process and continues until its assured that all the critical paths within the Block meets timing. Paths between Blocks are also analyzed.
7	Develop behavioral model (i.e RTL model) for each of the Sub-Blocks (i.e Megacell, Control and Datapath) within the Block. Even though here I have shown the Block to be having 1 Megacell, 1 Datapath and 1 Control, In reality a Block could have multiple Megacell subblocks, multiple Datapath subblocks and multiple Control subblocks. The sole purpose for doing so was to simplify the flowchart for better description.
8	Generally a Block gets verified for functional bugs in 3 different environments which are Stand Alone Test Bench (SAT) environment, Formal verification environment and Full Chip verification environment. **Full Chip** - Build a chip level RTL/Gate/Mixed (i.e a mixture of RTL and Gate netlists for the Blocks/Subblocks (we use this when Gate netlists for some of the Blocks/Subblocks is not yet ready or in some cases for case in debug)) model and verify it in a Full Chip simulation environment. The tests run in this environment are Directed tests and Pseudo Random tests generated by Pseudo Random test generator. Directed assembly tests/diags are manually written tests to focus on a particular functional aspect of a design. These tests could be self-checking or non-self-checking. In the case of self-checking tests they have built in checkers which compare the observed data against the expected ones and throws an error in the case of a mismatch. In the case of non-self-checking tests, they are run simultaneously on both the Chip model and the golden ISS (Instruction Set Simulator (Instruction Accurate Architectural

Model)) and a stand alone checker checks the architectural state (i.e various architectural registers) in both the Chip model as well as ISS on instruction boundaries. If there is a mismatch then the checker throws an error. Pseudo Random test generator in many cases is a internally developed test generator whose main responsibility is to generate random assembly diags based on a template or weightage provided by the user. These random diags in most cases are non self checking diags and are run simultaneously on both the Chip and golden ISS model to find any bugs. Any bugs found in this environment are fixed in the RTL/Gate netlists for the Blocks/Subblocks and the process of rebuilding the RTL/Gate/Mixed model for the Chip and running the tests continues until all the bugs are found with very good Functional and Code coverage. Functional coverage here refers to the percentage of functionality verified and Code coverage refers to the percentage of Code covered. Generally functional coverage objects (written in Vera/Verilog/VHDL) are manually written to measure functional coverage whereas Code coverage is measured by the various software simulator tools (i.e VCS) or commercially available tools (i.e HDLScore). RN in the flowchart refers to RTL netlist and GN refers to Gate level netlist. In some cases to speed things up the RTL or Gate netlists of the entire Chip or portions of the Chip which take too many simulation cycles are mapped into an Hardware accelerator or Emulator before running the tests.

SAT - Build an RTL/Gate/Mixed (i.e a mixture of RTL and Gate netlists for the Subblocks (we use this when Gate netlists for some of the Datapath/Control Subblocks is not yet ready or in some cases for case in debug)) model for the Block (i.e Block 1 in this case) and verify it in a Stand Alone Test bench environment (i.e SAT). The tests run in this environment could be any of the following - Simple binary test vectors, Directed assembly diags (manually written assembly diags) or Assembly Diags generated from Pseudo Random Test generator. Simple binary test vector is a mixed sequence of 0's and 1's. Directed diags here could be self checking or non-self checking. In the case of non-self checking directed diags and pseudo random diags, they are run simultaneously both on the RTL/Gate/Mixed model and the golden ISS to find bugs in the design. Any bugs found in this environment are fixed in the RTL/Gate netlists and the process of rebuilding the RTL/Gate/Mixed model for the Block and running the tests continues until all the bugs are found with very good Functional and Code coverage. One of the main advantages of verifying a block in this environment is the easiness in setting up test cases.

Formal - Build an RTL model for the Block (i.e Block1 in this case) and run Formal verification tools to prove or disprove the Assertions built in the RTL. In the case of disproof the formal verification tool provides a counter example (i.e test vectors which result in disproving the Assertion) for debug purposes. Any bugs found here are fixed in the RTL and the process of rebuilding the RTL model for the Block and running formal verification tools on it continues until all the assertions are proved.

| 9 | **Control** subblock within a Block (i.e Block1 in this case) goes through the following steps -

1.Synthesize the block with proper timing constraints and generate a gate level netlist. Initially the block is synthesized with a default wire load model but once we have layout for the block custom wire load model is generated and is used in all future synthesis work for the block. Every time a block (i.e control block) goes through layout changes a new custom wire load model is generated and used.

2. Push the block through Floorplan and timing driven Place and Route tools to generate a layout database. LVS (layout versus schematic), DRC (design rule checking) and ERC (electrical rule checking) are run on the layout database to see if there are any errors. In case of errors the layout is fixed (could be custom fixes or pushing through the flow again) and the Physical Verification Checks (i.e LVS, DRC and ERC) are rerun. This is an iterative process and |

continues until the layout is LVS, DRC and ERC clean.

3. Run Mintime(to make sure there are no mintime paths within the control block), EM/IR (to make sure the current density and voltage drop are within the Spec), Noise (to make sure the noise induced by the aggressors is within the noise limit) and Clock flow (to make sure the clock network within the block meets the Min and Max Clock Spec) on the layout database to see if they meet the required Spec. If they don't meet the required Spec then appropriate edits are made to the layout database by pushing the block through the flow again or by custom edits and the process of rerunning the above mentioned checks continues until all the checks meet the required Spec. Whenever the layout changes, the checks are rerun.

4. Extract Gate Level netlist and Parasitics. Run equivalence checking between the extracted gate level netlist and the RTL model to make sure that they are equivalent. Run Static Timing Analysis on the block with the extracted parasitics to see if it meets the required timing Spec. If it doesn't meet the timing Spec then branch to either **A3, B3, C3,** or **D3** as shown in the flowchart. Branch to **A3** if the path cannot be solved by simply resynthesizing the block, branch to **B3** if the path cannot be solved by incremental ECO, branch to **C3** if the path can be solved by incremental ECO or branch to **D3** if the database is frozen. Incremental ECO (Engineering Change Order) is a technique where the Place and Route tool updates the layout database with the required edits by retaining most of the previous placement and routing information, "database is frozen" refers to a stage in a design phase where the design is frozen (i.e no RTL changes and no pushing the block through P&R tools) and future edits need to be manually done in the layout.

| 10 | **Datapath** subblock within a Block (i.e Block1 in this case) goes through the following steps - |

1. Develop Structural code for the Behavioral RTL. Run equivalency checking to make sure that the Structural code is functionally equivalent to the behavioral RTL.

2. Develop a Placement file which provides the placement information (i.e Row and Column information) for each of the instances in the Structural code. The Placement tool uses this information to appropriately place the gates referred in the Structural code.

3. Once the Placement file is ready push the block through Floorplan and timing driven Place and Route tools to generate a layout database. LVS, DRC and ERC are run on the layout database to see if there are any errors. In case of any errors the layout is fixed (could be custom fixes or pushing it through the flow again) and the Physical Verification Checks are rerun. This is an iterative process and continues until the layout is LVS, DRC and ERC clean.

4. Run Min time, EM/IR, Noise and Clock flow on the layout database to see if they meet the required Spec. If they don't meet the required Spec then appropriate edits are made to the layout database by pushing the block through the flow again or by custom edits and the process of rerunning the above mentioned checks continues until all the checks meet the required Spec. Whenever the layout changes, the checks are rerun.

5. Extract Gate Level netlist and Parasitics. Run equivalence checking between the extracted gate level netlist and the RTL model to make sure that they are equivalent. Run Static Timing Analysis on the block with the extracted parasitics to see if it meets the required timing Spec. If it doesn't meet the timing Spec then branch to either **A2, B2, C2,** or **D2** as shown in the flowchart. Branch to **A2** if the path cannot be solved by simply updating the Structural code for the block, branch to **B2** if the path cannot be solved by incremental ECO, branch to **C2** if the path can be solved by incremental ECO or branch to **D2** if the database is frozen.

11	**Megacell** subblock within a Block (i.e Block 1 in this case) goes through the following steps - 1. Capture Schematics for the behavioral RTL using one of the Schematic editor tools. 2. Extract Transistor Level netlist and run equivalency checking to make sure that the transistor level netlist is functionally equivalent to its corresponding behavioral RTL. 3. Custom floorplan and draw Polygons for the transistors in the Schematics using one of the custom Layout Editor tools. Route the polygons manually. LVS, DRC and ERC are run on the layout database to see if there are any errors. In case of errors the layout is fixed and the Physical Verification Checks are rerun. This is an iterative process and continues until the layout is LVS, DRC and ERC clean. 4. Run Min time, EM/IR, Noise and Clock flow on the layout database to see if they meet the required Spec. If they don't meet the required Spec then appropriate edits are made to the layout database and the process of rerunning the above mentioned checks continues until all the checks meet the required Spec. Whenever the layout changes, the checks are rerun. 5. Extract Transistor Level netlist from the layout database and run Switch-Level Simulation using Playback vectors to see if there are any bugs in the netlist. If there are bugs then update Schematics and go through **C1** (as shown in the Flowchart) again. Re-extract the Transistor Level netlist once the layout is LVS, DRC and ERC clean and rerun Switch-Level Simulation. This process continues until the Playback vectors run bug free. Playback vectors are test vectors which provides both the input test patterns and the expected data for each of the input test patterns. 6. Extract Parasitics and run Dynamic Timing Analysis on the block to see if it meets the required timing Spec. If it doesn't meet the timing Spec then branch to either **A1**, **B1** or **C1** as shown in the flowchart. Branch to **A1** if the path cannot be solved by simply updating the Schematics, branch to **B1** if the path cannot be solved by editing the layout or branch to **C1** if the path can be solved by editing the layout.
12	Run Block level Static Timing analysis by using the extracted Parasitics for Datapath and Control blocks and Black box model for the Megacell. If the timing does not meet the required timing Spec then branch to either **A1, A2, A3, B1, B2, B3, C1, C2, C3, D2** or **D3** as shown in the flowchart to fix timing. Fixing timing in one or more branches and rerunning Static Timing on the block continues until the timing Spec is met.
13	Floorplan the Chip using one of the custom Floorplan tools and Place and Route Blocks/Subblocks within the Chip. In many cases most of the routes between blocks are shielded. Run LVS, DRC and ERC on the layout database and fix any errors in the layout. Run Min time, EM/IR, Noise and Clock flow on the Chip layout database to see if they meet the required Spec. If they don't meet the required Spec then appropriate edits are made to the layout database and the process of rerunning the above mentioned checks continues until all the checks meet the required Spec. Whenever the layout changes, the checks arc rerun. Extract Parasitics and run Static Timing Analysis on the Chip to see if it meets the required timing Spec. If it doesn't meet the timing Spec then branch to either **A1, A2, A3, B1, B2, B3, C1, C2, C3, D2, D3** or **F** as shown in the flowchart. Fixing timing in one or more branches and rerunning Static Timing on the Chip continues until the timing spec is met.
14	When all "**Done**'s" in the flowchart are satisfied, **Tapeout** the Chip.

Figure below shows the flowchart for a typical Processor design flow.

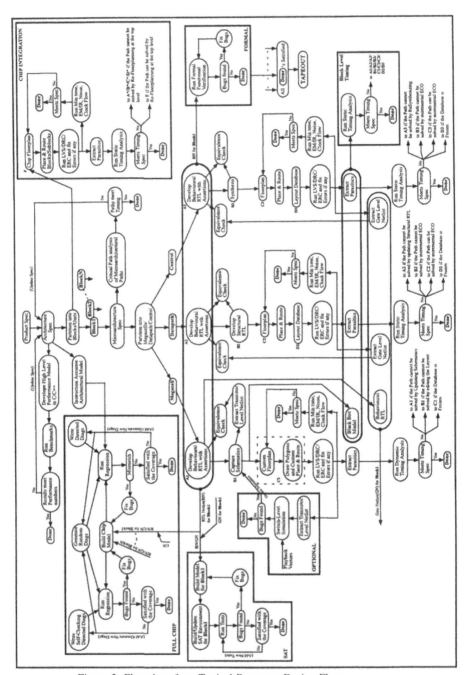

Figure 2: Flowchart for a Typical Processor Design Flow

4. What is the difference between a CISC and a RISC Processor?

CISC (Complex Instruction Set Computer) processors use complex instruction set whereas RISC (Reduced Instruction Set Computer) processors use reduced instruction set. Additional characteristics of CISC and RISC Processors are tabulated below.

Table 2: Comparison of CISC against RISC

CISC	RISC
1. Variable instruction length.	1. Fixed instruction length.
2. Large number of addressing modes.	2. Few addressing modes.
3. Support for small number of general purpose registers.	3. Support for large number of general purpose registers.
4. Requires less number of instructions to represent an application code when compared to RISC.	4. Requires more number of instructions to represent an application code when compared to CISC although this is debatable.
5. Requires complex Compiler.	5. Requires less complex Compiler.
6. Since the application code compiled for CISC instruction set results in less number of instructions we need less memory to store the application binaries in a CISC machine.	6. Since the application code compiled for RISC instruction set results in more number of instructions (when compared to CISC) we need more memory to store the application binaries in a RISC machine.
7. Less number of instructions need not necessarily mean that an application running on a CISC processor results in higher performance than the same running on a RISC processor.	7. More number of instructions need not necessarily mean that an application running on a RISC processor results in lower performance than the same running on a CISC processor.
8. In addition to Load/Store there are other instructions which results in accessing memory.	8. Load/Store (Atomics included) are the only ones which can access memory.
9. A typical CISC instruction *(Intel x 86 instruction)* Assembly Syntax ADD AL, BL (AL + BL -> AL)	9. A typical RISC instruction *(SPARC instruction)* Assembly Syntax ADD rs1, rs2, rd (rs1 + rs2 -> rd)
10. Examples of CISC processors are Intel's 486, Pentium (all flavours), AMD's Krypton, Athlon etc.	10. Examples of RISC processors are SUN's UltraSparc, MIPS's MIPS32, MIPS64, ARM'S ARM11, Motorola's PowerPC etc.

5. What is Cache Memory in a CPU and what are the most common terms associated with Caches?

A Cache is a small high speed memory which stores the most recently used instructions or data from a larger but slower memory system. A Cache could be residing on-chip or off-chip and there could be more than one level of cache in a memory hierarchy. In Figure 3 below, fig (a) has 2 levels of caches, one on-chip (assuming here that we have a separate Level 1 cache for Instruction and Data) and one off-chip, fig (b) has three levels of caches, two on-chip and one off-chip, fig (c) has four levels of caches, three on-chip and one off-chip.

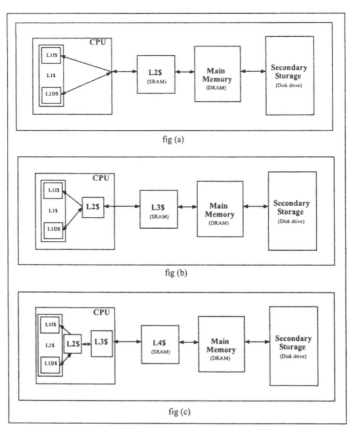

Figure 3: Various Cache Levels

Table below shows the most common terms associated with Caches.

Table 3: Common Terms Associated with Caches

Term	Description
Bit, Nibble, Byte, HalfWord, Word, Double Word, Quad-Word, Line, Page	*Bit* - It is a group of 1 -bit, *Nibble* - It is a group of 4-bits, *Byte* - It is a group of 8-bits, *HalfWord* - It is a group of 16 bits, *Word* - It is a group of 32-bits, *DoubleWord* - It is a group of 64-bits, *QuadWord* - It is a group of 128-bits, *Line* - It is the unit of transfer between Main Memory and Cache or between a higher level Cache and a lower level Cache, *Page* - It is the unit of transfer between Main Memory and secondary storage (i.e Hard Drive or Tape etc.) Figure below shows the definition of the above mentioned terms as applied to Main Memory having several Pages. 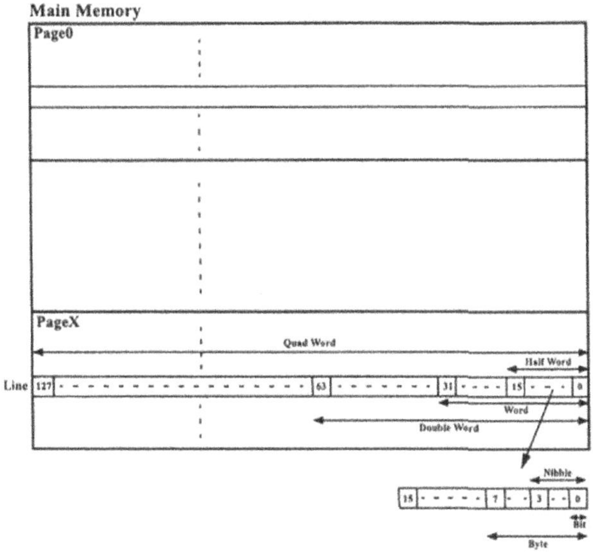 Figure 4: Bit, Nibble, Byte, HalfWord, Word, DoubleWord, QuadWord, Line and Page
Cold Start, Warm Start	These are the terms used for Cache Performance evaluation. Cold start is a condition where simulation assumes no instructions or data corresponding to the benchmark or application sitting in the Cache. Warm start is a condition where simulation assumes instructions or data corresponding to the benchmark or application sitting in the Cache.

Table 3: Common Terms Associated with Caches

Term	Description
Hit, Miss	Hit is a condition where the CPU finds the required data in Cache whereas Miss is a condition where the CPU doesn't find the required data in the Cache. Figure below shows the Cache Hit and Miss condition. 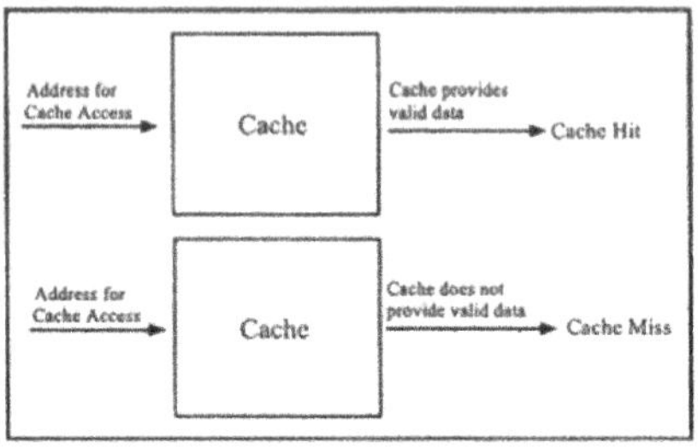 Figure 5: Cache Hit and Cache Miss
Temporal Locality, Spatial Locality, Sequential Locality	*Temporal Locality* - If a location is referenced then it is likely to be referenced again in the near future. *Spatial Locality* - If a location is referenced then it is likely that locations near it will be referenced in the near future. *Sequential Locality* - This is a special case of Spatial locality where the address of the next reference will be the immediate successor of the present reference. Figure below shows Temporal, Spatial and Sequential locality. 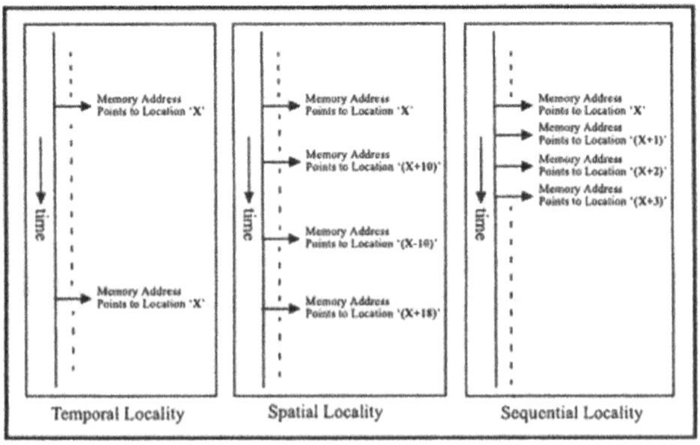 Figure 6: Temporal Locality, Spatial Locality and Sequential Locality

Table 3: Common Terms Associated with Caches

Term	Description
Compulsory Misses, Capacity Misses, Conflict Misses	Compulsory misses are due to programs first reference to a memory block. These misses can not be prevented by any caching technique. Capacity misses are due to insufficient capacity in a Cache. These misses can be prevented to a certain extent by increasing the size of the Cache. Conflict misses are due to insufficient associativity in a direct mapped or a set associative Cache implementation. These misses can be prevented to a certain extent by increasing the associativity of the Cache.
Virtually Indexed Physically Tagged, *Physically Indexed Physically Tagged*	A Cache indexed by a Virtual Address and tagged with a Physical Address is known as a Virtually Indexed Physically Tagged (i.e VIPT) Cache and a Cache indexed by a Physical Address and tagged with a Physical Address is known as a Physically Indexed Physically Tagged (i.e PIPT) Cache. One of the problems with VIPT caches is address aliasing i.e two Virtual Addresses getting mapped to the same Physical Address. Using page offset (since page offset is unaffected by address translation) to index the Cache results in avoiding any address aliasing issues with VIPT Caches. Figure below shows VIPT and PIPT Caches. 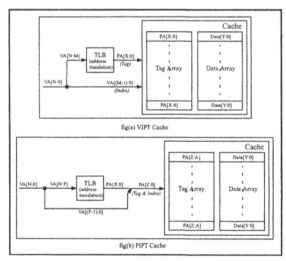 Figure 7: VIPT and PIPT Caches

Table 3: Common Terms Associated with Caches

Term	Description
Load use Latency	It is the number of cycles a Load dependent instruction has to wait before it can be issued for execution from the time the Load gets issued to the address generation logic. Typically the load use latency for a Load that hits in Level 1 Cache is 3 to 4 cycles and for the one that Misses in Level 1 Cache is 8 to 15 cycles. Figure below shows load use latency. 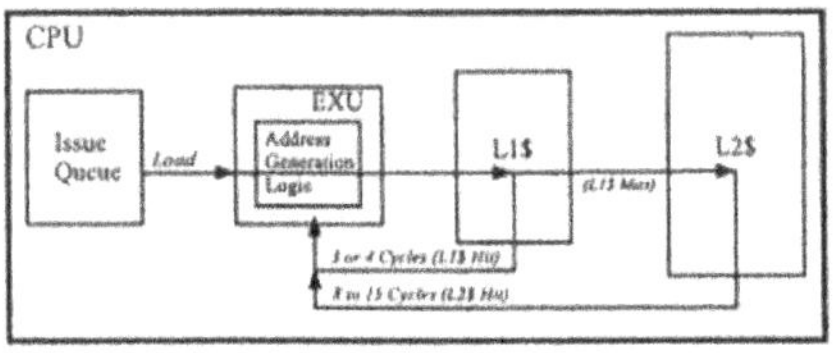 Figure 8: Load Use Latency
Harvard Architecture, *Berkeley Architecture*	Harvard Architecture is a split architecture where you have a separate dedicated Cache for Instructions and Data. Berkeley Architecture is a unified architecture where you have one unified Cache for both Instructions and Data. Figure below shows Harvard and Berkeley architecture as applied to a Level 1 Cache. 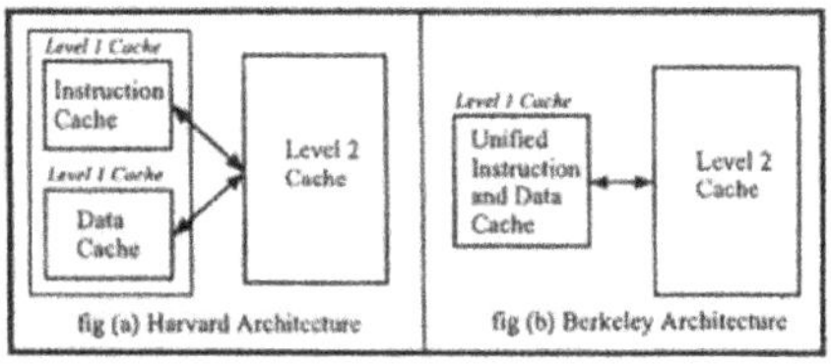 Figure 9: Harvard and Berkeley Architecture
Store Allocate, *Store Non-Allocate*	Store Allocate is a write miss policy where the block (line) is loaded from either higher level cache or main memory in the case of a write miss in lower level cache. Store Non-Allocate is a write miss policy where the block is modified in higher level cache but not allocated in lower level cache in the case of a write miss in lower level cache. Figure below shows Store Allocate and Store Non-Allocate cache policies.

Table 3: Common Terms Associated with Caches

Term	Description
	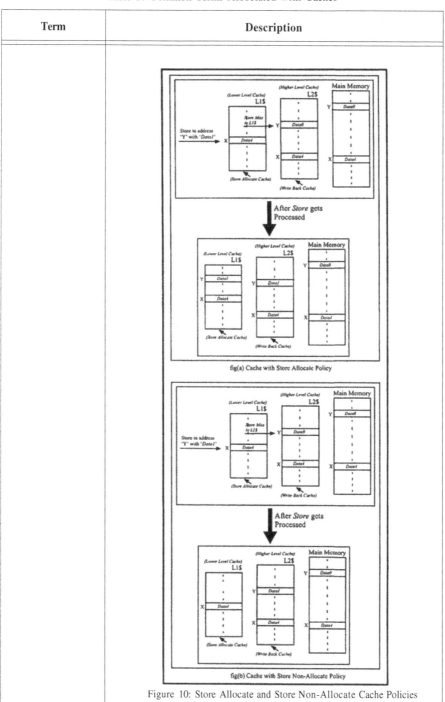

Figure 10: Store Allocate and Store Non-Allocate Cache Policies

Table 3: Common Terms Associated with Caches

<table>
<tr><th>Term</th><th>Description</th></tr>
<tr><td></td><td></td></tr>
<tr><td>Write Back,
Write Through</td><td>
Write Back is a write policy where the corresponding memory block is updated only when the block in cache is selected for replacement.

Write Through is a write policy where the corresponding memory block is updated whenever the block in cache is updated or the corresponding cache block in higher level cache maintaining cache coherency is updated whenever the block in lower level cache is updated.

Figure below shows Write Back and Write Through policies.

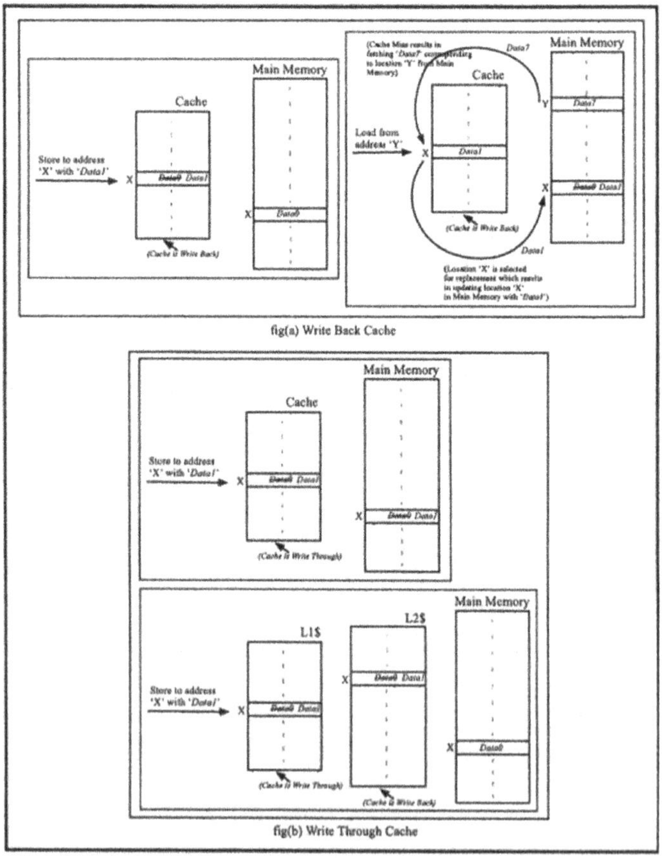

fig(a) Write Back Cache

fig(b) Write Through Cache

Figure 11 : Write Back and Write Through Cache
</td></tr>
<tr><td></td><td></td></tr>
</table>

Table 3: Common Terms Associated with Caches

Term	Description
Multi-Ported, *Multi-Banked,* *Multi-Pumped*	*Multi-Ported Cache* It is a cache implementation where the cache provides support for more than one Read or Write port for providing high bandwidth. Because of these multiple ports it results in servicing multiple requests per cycle (i.e a 2 Read port cache can service two read requests per cycle). A 2 port, 4-Way Set Associative cache is shown in Figure 12. We see from the figure that irrespective of the type of address we result in servicing two requests in any given cycle. 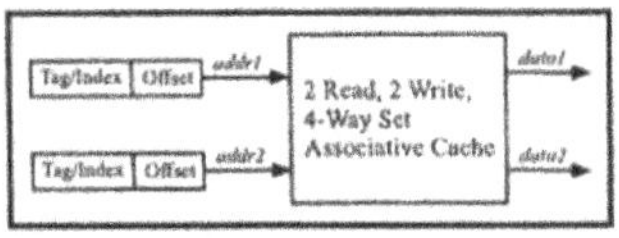 Figure 12: Multi-Ported Cache Typically this kind of implementation gets used in lower level caches (i.e level 1) where the cache size is small. *Multi-Banked Cache* It is a cache implementation where the cache is implemented as a banked structure for providing high bandwidth by providing the illusion of multiple ports. Here it results in servicing multiple requests per cycle if there are no bank conflicts. A 4-Way Set Associative banked cache is shown in Figure 13 below. We see from the figure that it has 4 banks with each bank having all 4 Ways. Here we can process all 4 requests if there is no bank select conflict between the requested addresses. 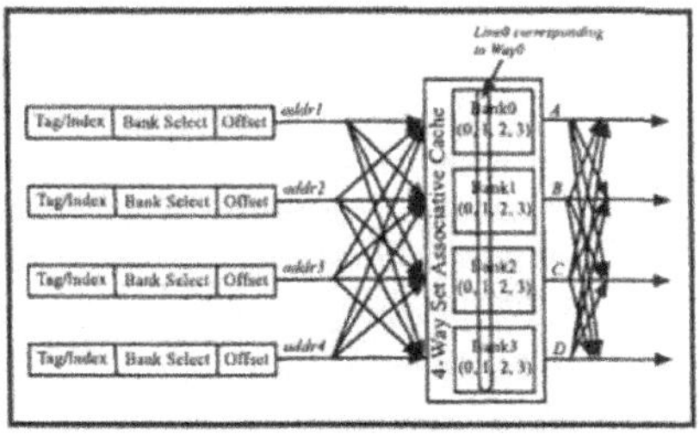 Figure 13: Multi-Banked Cache Typically this kind of implementation gets used in higher level caches (i.e level 2, level 3 etc.) where the cache size is big.

Table 3: Common Terms Associated with Caches

Term	Description
	Multi-Pumped Cache It is a cache implementation where you time multiplex a single port thereby providing multiple accesses in a given cycle. Here the cache is superpipelined (i.e cache is operating at a higher frequency compared to the rest of the pipe) for providing high bandwidth. A double pumped 4-Way Set Associative cache is shown in Figure 14 below. We see from the figure that by double pumping we can service two requests in a given cycle. 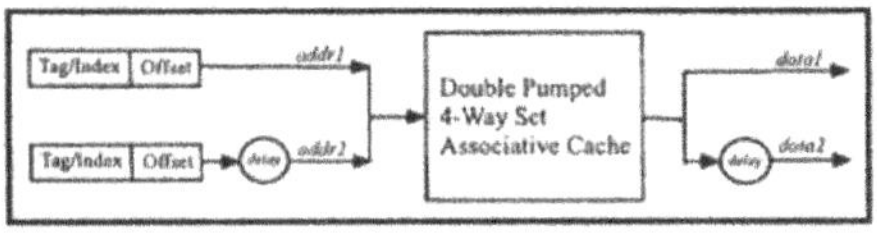 Figure 14: Multi-Pumped Cache
Inclusive, Non-Inclusive	Inclusion is a cache property where the contents of lower level cache are a subset of higher level cache i.e in a two level cache hierarchy, all the contents of Level 1 cache are a subset of Level 2 cache (i.e a Hit in Level 1 guarantees a Hit in Level 2 and a Miss in Level 1 need not necessarily be a Miss in Level 2). Maintaining inclusion in a two level cache hierarchy system is fairly straight forward. Whenever a block or line enters Level 1 cache it must also be placed in Level 2 cache and, whenever a line leaves Level 2 cache (because of eviction due to line replacement) or is invalidated in Level 2 cache (because of external snoop) it must also be invalidated in Level 1 cache if the block/line happens to reside in Level 1 cache. Non-Inclusion is a cache property where the contents of lower level cache need not necessarily be a subset of higher level cache i.e in a two level cache hierarchy, the contents of Level 1 cache need not necessarily be sitting in Level 2 cache (i.e a Hit/Miss in Level 1 cache need not necessarily be a Hit/Miss in level 2 cache). Maintaining non-inclusion in a two level cache hierarchy system requires the following: coherency is maintained independently for each cache level (i.e Level 1 and Level 2 here) i.e in a snooping implementation this implies that both Level 1 and Level 2 of the cache hierarchy must snoop all addresses over the system bus. Figure below shows Inclusive and Non-Inclusive cache property.

Table 3: Common Terms Associated with Caches

Term	Description
	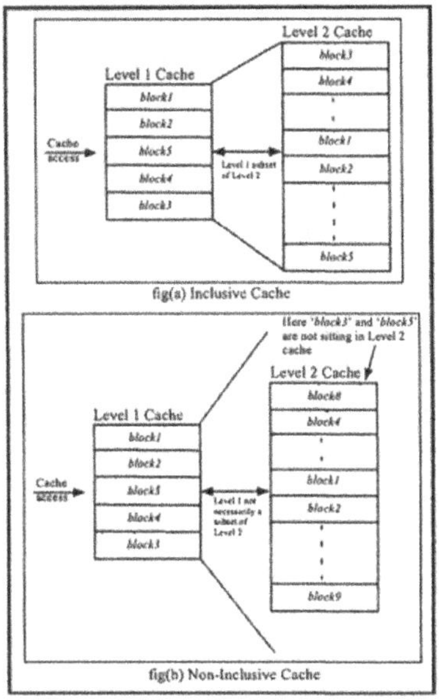 Figure 15: Inclusive and Non-Inclusive Cache
Cacheable, *Non-Cacheable*	Data that can be cached (i.e written into cache) is known as Cacheable data whereas data that cannot be cached (i.e can not be written into cache) is known as Non-Cacheable data. Figure below shows Non-Cacheable load access. 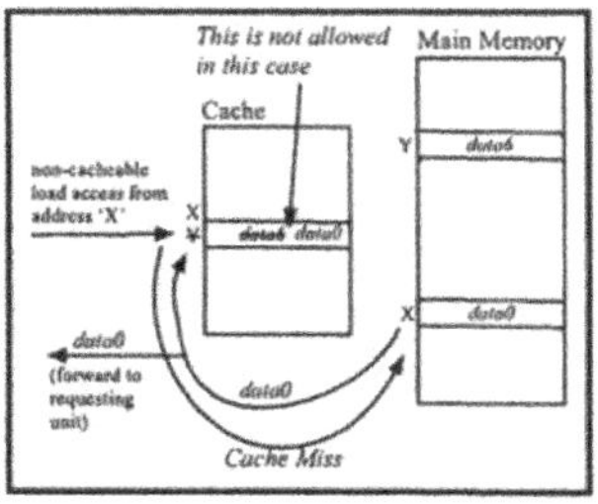 Figure 16: Non-Cacheable Load access

Table 3: Common Terms Associated with Caches

Term	Description
Big-Endian, Little-Endian	Big-Endian and Little-Endian are terms that describe the order in which a sequence of bytes are stored in a Cache. Big-Endian is an addressing convention where the byte with the smallest address is the most significant byte in a multi-byte word. Little-Endian is an addressing convention where the byte with the smallest address is the least significant byte in a multi-byte word. Figure 17 below shows the way data gets read and written into a direct-mapped Data Cache in the case of a Big-Endian and Little-Endian addressing convention. This is true for any Cache configuration. We see from the figure that while the Cache is supporting Big-Endian addressing convention the most significant byte (i.e B7) of the incoming 8 byte word gets written at the lowest address location (i.e m000) and the least significant byte (i.e B0) gets written at the highest address location (i.e m111) whereas, while the Cache is supporting Little-Endian addressing convention we see that the least significant byte (i.e B0) of the incoming 8 byte word gets written at the lowest address location (i.e m000) and the most significant byte (i.e B7) gets written at the highest address location (i.e m111). Processors supporting Big-Endian byte order by default are Sparc processors whereas processors supporting Little-Endian byte order by default are Intel's x86 processors, ARM processors etc. Figure 17: Big-Endian and Little-Endian

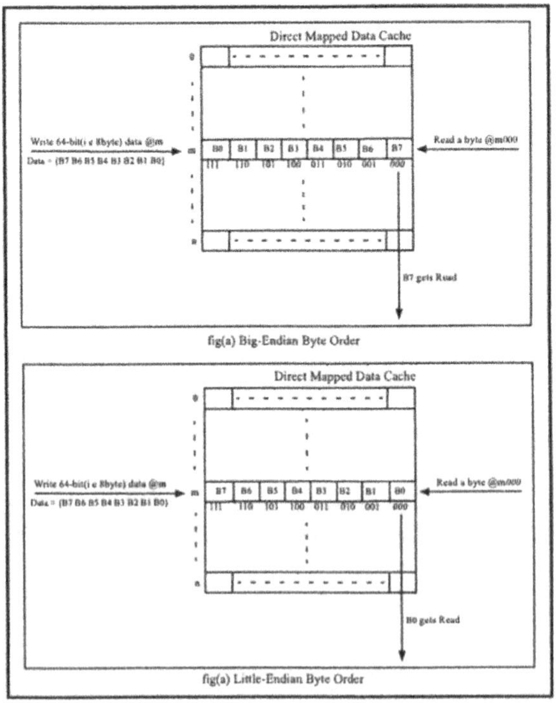

6. What are the various Cache mappings?

The various Cache Mappings are tabulated below.

Table 4: Cache Mappings

Cache Mappings	Description
Direct Mapped Cache	In this kind of cache a given main memory line can be placed in one and only one place in the cache.
Fully Associative Cache	In this kind of mapping any line from main memory can be placed anywhere in the cache.
Set Associative Cache	This is similar to direct mapped but in this case more than one line from each set reside in the cache.

7. Describe a Direct Mapped Cache Memory with an example?

Figures 18 and 19 below shows a Direct Mapped Cache Memory in the case of a Hit and a Miss. In Figures 18 and 19, the following things are assumed -

16-bit Virtual Address (VA, i.e the address seen by the Programmer)
16Kbyte (KB) Main Memory (MM)
8KB Page size
1B Line size (Unit of transfer between Main Memory and Cache)
256B L1 Cache
2-entry Fully Associative TLB (Translation Look Aside Buffer)

Following facts can be drawn from the assumptions made above -

Since the size of Main Memory is 16KB, we need a 14-bit Physical Address (PA, i.e the address to index any byte within the Main Memory).
Since the size of L1 Cache (L1$) is 256B and the Line size is 1B, L1 Data will hold 256 lines from Main Memory and L1Tag will hold 256 Tag address where each Tag address corresponds to a Line in L1Data.
Assuming that we would like to access one Line of data (i.e 1Byte in this case) on every access from L1$ and since L1$ is 256 entries (L1$ being Direct Mapped), we need 8-bits to index L1Data and L1Tag.

Since Main Memory is 16KB, we will have a maximum of 2 Pages (16KB/8KB=2) sitting in Main Memory at any given time. Each Page in Main Memory has 8K Lines (i.e 8KB(Page size)/1B(Line size)=8K Lines).

Since L1$ in this case has 256 entries, any access for a Miss in L1$ will see Main Memory to be having 256 Sets with each Set having 64 Lines each (16K Lines/256=64 Lines) as shown in Figures 18 and 19.

Since in a Direct Mapped Cache only one Line from each Set could be sitting in the Cache at any given point of time, one of the lines, LineA, LineC, LineW, LineY or any other Line belonging to Set 0 could be sitting in entry 0 of L1$, Similarly one of the lines, LineB, LineD, LineX, LineZ or any other Line belonging to Set 255 could be sitting in entry 255 of L1$.

Description for the Hit case

In Figure 18 we have assumed that LineA from Set 0 is sitting in entry 0 and LineD from Set 255 is sitting in entry 255 of the L1$. The program while it gets executed on CPU always generates a VA (i.e VA1 in this case) and this address needs to be translated to PA before we can access data from the Cache or Main Memory. TLB in this case translates the VA into PA. If we assume here that the VA (i.e VA1) generated by the program is 16'b0011 1111 1111 1111, TLB here only has to translate the most significant 3-bits of the VA into a Physical Page number. This is because since the Page size is 8KB, the least significant 13-bits of PA should be same as the least significant 13-bits of VA. Since only 2 Pages could be residing in Main Memory at any given point of time, TLB translates the upper 3-bits of VA to either 0 or 1 assuming that we don't have a Page Fault (a condition where the requested Page is not sitting in Main Memory). In Figure 18 below TLB translates Virtual Page Number 3'b001 (i.e most significant 3 bits of VA 16'b0011 1111 1111) to 1'b0 (Physical Page Number (i.e the most significant bit(s) of PA). Since L1$ has 256 entries and is Direct Mapped, the least significant 8-bits of the PA gets used to index both the Tag and Data portion of the Cache (Since here the least significant 13-bits of the PA is same as the least significant 13-bits of the VA we could as well use the least significant 8-bits of the VA to index the Tag and Data portion of the Cache to make things faster). Once accessed the data from Tag array gets compared against the Tag portion of the PA (i.e in this case the most significant 6-bits of PA1 which is 6'b01 1111). Since here we have a match, we have a Cache Hit and the data read from Data portion of the Cache gets forwarded to the Unit requesting it.

Description for the Miss case

In Figure 19 we have assumed that LineA from Set 0 is sitting in entry 0 and LineD from Set 255 is sitting in entry 255 of the L1$. If we assume here that the VA (i.e VA2) generated by the program is 16'b1011 1111 1111 1111 then the TLB translates the most significant 3-bits of the VA (i.e 3'b101, Virtual Page Number) to 1'b1 (i.e Physical Page Number). On Tag comparison after accessing the Tag array we find that it results in a mismatch thereby resulting in a Cache Miss. In such case the Cache requests data from Main Memory in which case LineZ from Main Memory replaces LineD in Data array and 6'b11 1111 replaces 6'b01 1111 in Tag array. LineZ also gets forwarded to the Unit requesting it.

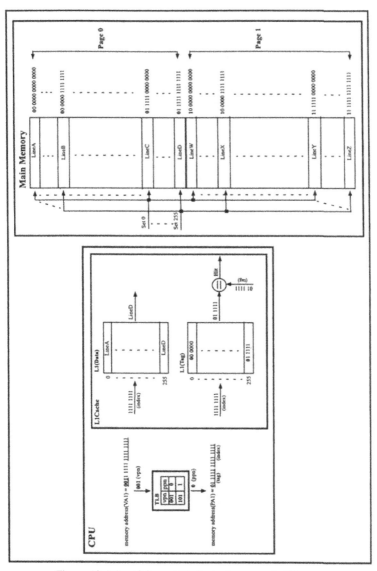

Figure 18: Direct Mapped Cache in the case of a Hit

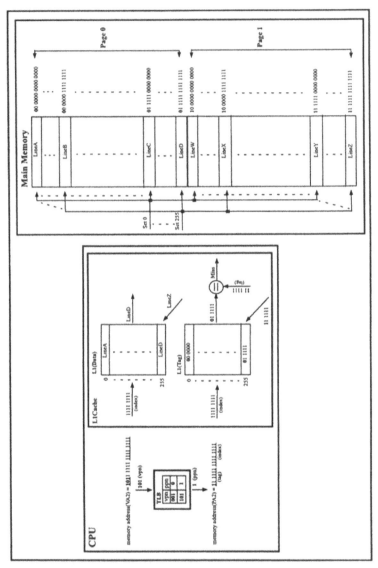

Figure 19: Direct Mapped Cache in the case of a Miss

8. Describe a Fully Associative Cache Memory with an example?

Figures 20 and 21 below shows a Fully Associative Cache Memory in the case of a Hit and a Miss. In Figures 20 and 21, the following things are assumed

| 16-bit Virtual Addres (VA, i.e the address seen by the Programmer) |
| 16Kbyte (KB) Main Memory (MM) |
| 8KB Page size |
| 1B Line size (Unit of transfer between Main Memory and Cache) |
| 256B L1 Cache |
| 2-entry Fully Associative TLB (Translation Look Aside Buffer) |

Following facts can be drawn from the assumsions made above -

Since the size of Main Memory is 16KB, we need a 14-bit Physical Address (PA, i.e the address to index any byte within the Main Memory).

Since the size of L1 Cache (L1$) is 256B and the Line size is 1B, L1 Data will hold 256 lines from Main Memory and L1Tag will hold 256 Tag address where each Tag address corresponds to a Line in L1 Data.

Since Main Memory is 16KB, we will have a maximum of 2 Pages (16KB/8KB=2) sitting in Main Memory at any given time. Each Page in Main Memory has 8K Lines (i.e 8KB(Page size)/1B(Line size)=8K Lines).

Since L1$ in this case has 256 entries, any access for a Miss in L1$ will see Main Memory to be having 256 Sets with each Set having 64 Lines each (16K Lines/256=64 Lines) as shown in Figures 20 and 21.

Since in a Fully Associative Cache any line from any Set could be sitting anywhere in the Cache, any of the lines (i.e LineA, LineB, LineC, LineD, LineW, LineX, LineY, LineZ etc.) in Main Memory could be sitting in entry0 of L1$. Similarly any of the lines in Main Memory could be sitting in entry 255 of L1$.

Description for the Hit case

In Figure 20 we have assumed that LineD from Set 255 is sitting in entry 0 and LineA from Set 0 is sitting in entry 255 of the L1$. The program while it gets executed on CPU always generates a VA (i.e VA1 in this case) and this address needs to be translated to PA before we can access data from the Cache or Main Memory. TLB in this case translates the VA into PA. If we assume here that the VA (i.e VA1) generated by the program is 16'b**0011** 1111 1111 1111, TLB here only has to translate the most significant 3-bits of the VA into a Physical Page number. This is because since the Page size is 8KB, the least significant 13-bits of PA should be same as the least significant 13-bits of VA. Since only 2 Pages could be residing in Main Memory at any given point of time, TLB translates the upper 3-bits of VA to either 0 or 1 assuming that we don't have a Page Fault (a condition where the requested Page is not sitting in Main Memory). In Figure 20 below TLB translates Virtual Page Number 3'b**001** (i.e most significant 3 bits of VA 16'b**0011** 1111 1111 1111) to 1'b**0** (Physical Page Number (i.e the most significant bit(s) of PA). Since L1$ has 256 entries and is Fully Associative, all the bits of PA are compared against all the entries of the Tag array resulting in a one-hot vector (i.e in the case of a Hit) which gets used to select one of the 256 entries of the Data array. Since here the PA matches with entry 0 of the Tag array, we have a Cache Hit. In this case data sitting in entry 0 (i.e LineD) of the Data array gets forwarded to the Unit requesting it.

Description for the Miss case

In Figure 21 we have assumed that LineD from Set 255 is sitting in entry 0 and LineA from Set 0 is sitting in entry 255 of the L1$. If we assume here that the VA (i.e VA2) generated by the program is 16'b**1011** 1111 1111 1111 then the TLB translates the most significant 3-bits of the VA (i.e 3 'b**101**, Virtual Page Number) to 1 'b**1** (i.e Physical Page Number). On Tag comparison against all the entries in Tag array we find that it results in a mismatch thereby resulting in a Miss. In such case the Cache requests data from Main Memory in which case LineZ from Main Memory replaces one of the entries in Data array (the Line getting replaced depends on the kind of replacement algorithm being used). LineZ also gets forwarded to the Unit requesting it.

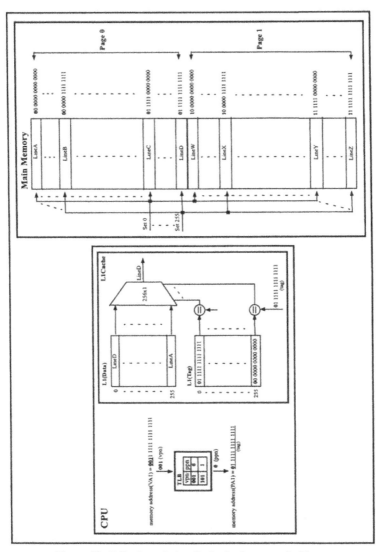

Figure 20: Fully Associative Cache in the case of a Hit

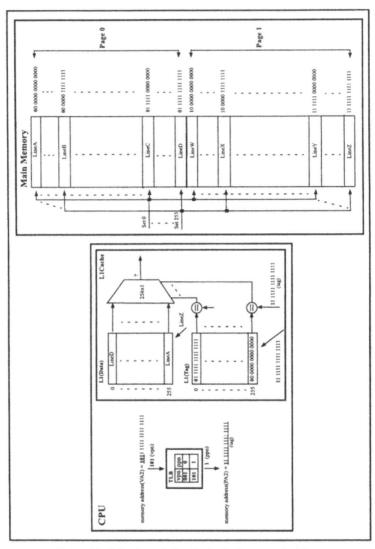

Figure 21: Fully Associative Cache in the case of a Miss

9. Describe a 2-Way Set Associative Cache Memory with an example?

Figures 22 and 23 below shows a 2-Way Set Associative Cache Memory in the case of a Hit and a Miss. In Figures 22 and 23, the following things are assumed -

16-bit Virtual Addres (VA, i.e the address seen by the Programmer)

16Kbyte (KB) Main Memory(MM)
8KB Page size
1B Line size (Unit of transfer between Main Memory and Cache)
256B L1 Cache
2-entry Fully Associative TLB (Translation Look Aside Buffer)

Following facts can be drawn from the assumsions made above -

Since the size of Main Memory is 16KB, we need a 14-bit Physical Address (PA, i.e the address to index any byte within the Main Memory).
Since the size of L1Cache (L1$) is 256B and the Line size is 1B, L1Data will hold 256 lines from Main Memory and L1Tag will hold 256 Tag address where each Tag address corresponds to a Line in L1Data.
Assuming that we would like to access one Line of data (i.e 1Byte in this case) on every access from L1$ and since L1$ is 128x2 entries (L1$ being 2-Way Set Associative), we need 7-bits to index L1Data and L1Tag arrays.
Since Main Memory is 16KB, we will have a maximum of 2 Pages (16KB/8KB=2) sitting in Main Memory at any given time. Each Page in Main Memory has 8K Lines (i.e 8KB(Page size)/1B(Line size)=8K Lines).
Since L1$ in this case has 128x2 entries, any access for a Miss in L1$ will see Main Memory to be having 128 Sets with each Set having 128 Lines each (16K Lines/128=128 Lines) as shown in Figures 22 and 23.
Since in a 2-Way Set Associative Cache two Lines from each Set could be sitting in the Cache at any given point of time, one of the lines, LineA, LineC, LineW, LineY or any other Line belonging to Set 0 could be sitting in entry 0 of Way0 and Way1 of L1$. Similarly one of the lines, LineB, LineD, LineX, LineZ or any other Line belonging to Set 127 could be sitting in entry 127 of Way0 and Way1 of L1$.

Description for the Hit case

In Figure 22 we have assumed that LineA from Set 0 is sitting in entry 0 of Way0, LineC from Set 0 is sitting in entry 0 of Way1, LineD from Set 127 is sitting in entry 127 of Way0 and LineX from Set 127 is sitting in entry 127 of Way1. The program while it gets executed on CPU always generates a VA (i.e VA1 in this case) and this address needs to be translated to PA before we can access data from the Cache or Main Memory. TLB in this case translates the VA into PA. If we assume here that the VA (i.e VA1) generated by the program is 16'b0011 1111 1111 1111, TLB here only has to translate the most significant 3-bits of the VA into a Physical Page number. This is because since the Page size is 8KB, the least significant 13-bits of PA should be same as the least significant 13-bits of VA. Since only 2 Pages could be residing in Main Memory at any given point of time, TLB translates the upper 3-bits of VA to either 0 or 1 assuming that we don't have a Page Fault (a condition where the requested Page is not sitting in Main Memory). In Figure 22 below TLB translates Virtual Page Number 3'b001 (i.e most significant 3 bits of VA 16'b0011 1111 1111 1111) to 1'b0 (Physical Page Number (i.e the most significant bit(s) of PA). Since L1$ has 128x2 entries and is 2-Way Set Associative, the least significant 7-bits of the PA gets used to index both the Tag and Data portion of Way0 and Way1 of the Cache (Since here the least significant 13-bits of the PA is same as the least significant 13-bits of the VA we could as well use the least significant 7-bits of the VA to index the Tag and Data portion of the Cache). Once accessed the data from Tag array of Way0 and Way1 gets compared against the Tag portion of the PA (i.e in this case the most significant 7-bits of PA1 which is 7'b01 1111). Since here we have a match as the tag matches with the one sitting in Tag array of Way0, we have a Cache Hit and the data read from Data portion of Way0 (i.e LineD) gets forwarded to the Unit requesting it.

Description for the Miss case

In Figure 23 we have assumed that LineA from Set 0 is sitting in entry 0 of Way0, LineC from Set 0 is sitting in entry 0 of Way1, LineD from Set 127 is sitting in entry 127 of Way0 and LineX from Set 127 is sitting in entry 127 of Way1. If we assume here that the VA (i.e VA2) generated by the program is 16'b1011 1111 1111 1111 then the TLB translates the most significant 3-bits of the VA (i.e e3'b101, Virtual Page Number) to 1'b1 (i.e Physical Page Number). On Tag comparison after accessing the Tag array of Way0 and Way1 we find that it results in a mismatch thereby resulting in a Cache Miss. In such case the Cache requests data from Main Memory in which case LineZ from Main Memory replaces LineD/LineX in Way0/Way1 (the Way that gets replaced depends on the kind of replacement algorithm being used). LineZ also gets forwarded to the Unit requesting it.

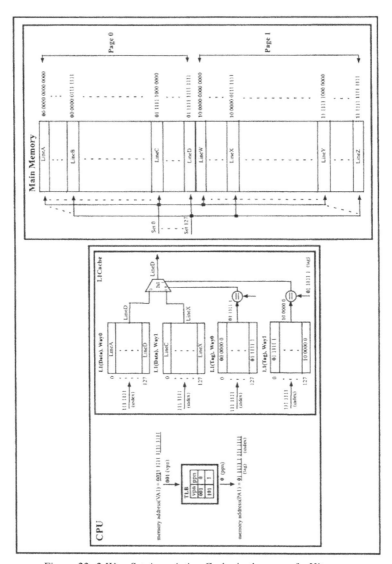

Figure 22: 2-Way Set Associative Cache in the case of a Hit

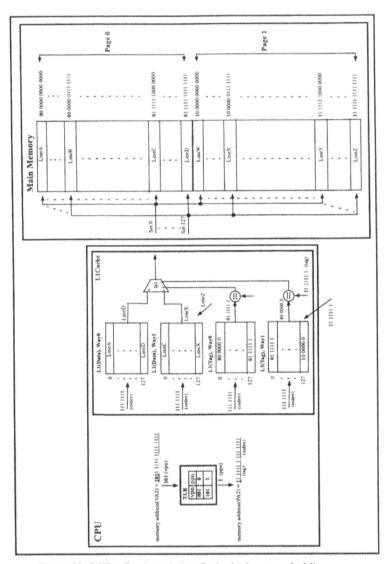

Figure 23: 2-Way Set Associative Cache in the case of a Miss

10. What are the most common Replacement Algorithms used in a design ?

The most common Replacement Algorithms used in a design are tabulated below.

Table 5: Replacement Algorithms

Algorithm	Description
Pseudo Random	Under this policy the replacement of a line is determined in a pseudo random fashion.
Full Random	Under this policy the replacement of a line is determined in a fully random fashion.
Pseudo LRU	Under this policy the line that was least/last recently used (need not necessarily be a true least/last recently used line) would be the candidate for replacement.
Full LRU	Under this policy the line that was least/last recently used (is a true least/last recently used line) would be the candidate for replacement.
Round Robin	Under this policy the replacement of a line in a cache happens in a round robin fashion.
MRU	Under this policy the line that was most recently used would be the candidate for replacement.
FIFO	Under this policy the line that had been in the cache for longest time would be the candidate for replacement.

11. Describe the way Pseudo Random Algorithm gets used in replacing an entry in a 4-Way Set Associative Cache Memory?

Table below shows the way Pseudo Random Algorithm gets used in replacing an entry in a 4-Way Set Associative Cache Memory.

Table 6: Pseudo Random Algorithm as Applied to a 4-Way Set Associative Cache Memory

1. Maintain a 2-bit Random array with number of entries equal to the number of entries in the Tag array as shown in Figure 24 below. Also maintain a LFSR (Linear Feedback Shift Register) whose output gets used to update the Random array as shown in figure below. Instead of an LFSR, a N-bit Up/Down Counter could also be used to update the Random array in which case the LFSR gets replaced by a N-bit Counter. LFSR gets updated every clock cycle. The feedback configuration for LFSR is chosen to provide a maximal length LFSR. In figure below the bits used to index the Data and Tag array may be same or different (i.e index0 = index1 or index0 != index).

Table 6: Pseudo Random Algorithm as Applied to a 4-Way Set Associative Cache Memory

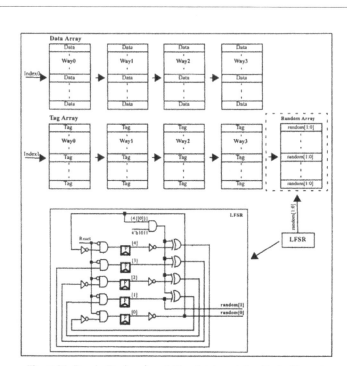

Figure 24: Pseudo Random for a 4-Way Set Associative Cache Memory

2. At Power on, reset the entire Random Array to all 0's.

3. Random array remains untouched (i.e doesn't get updated) in the case of a Cache Hit.

4. Random array remains untouched in the case of a Snoop Invalidate.

5. Use the following algorithm to replace an entry and update the Random array in the case of a Cache Miss.

```
if(random[1:0] for the indexed entry is 2'b00)
    Replace the indexed entry in Way0
    Set random[1:0] with the value from LFSR while you are updating the entry in Way0
elseif(random[1:0] for the indexed entry is 2'b01)
    Replace the indexed entry in Way1
    Set random[1:0] with the value from LFSR while you are updating the entry in Way1
elseif(random[1:0] for the indexed entry is 2'b10)
    Replace the indexed entry in Way2
    Set random[1:0] with the value from LFSR while you are updating the entry in Way2
else(random[1:0] for the indexed entry is 2'b11)
    Replace the indexed entry in Way3
    Set random[1:0] with the value from LFSR while you are updating the entry in Way3
```

12. Describe the way Full Random Algorithm gets used in replacing an entry in a 4-Way Set Associative Cache Memory?

Table below shows the way Full Random Algorithm gets used in replacing an entry in a 4-Way Set Associative Cache Memory.

Table 7: Full Random Algorithm as Applied to a 4-Way Set Associative Cache Memory

1. Maintain a 2-bit Random array with number of entries equal to the number of entries in the Tag array as shown in Figure 25 below. A 2-bit Random Number Generator Logic (RNGL) is used to generate the 2-bit Random vector which gets used to update the Random array as shown in figure below. RNGL makes use of two groups of control signals (i.e GroupA and GroupB) along with a even parity generator and a noise source to generate a random number as a result of which the output of RNGL is totally unpredictable (i.e Fully Random). These control signals could be tapped from anywhere within the chip. There could be other means of generating a Fully Random number. Random values gets updated every cycle. In figure below the bits used to index the Data and Tag array may be same or different (i.e index0 = index1 or index0 != index1).

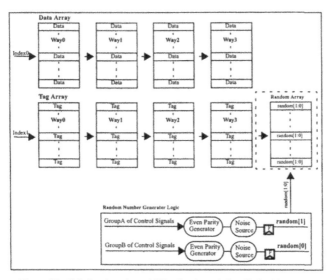

Figure 25: Full Random for a 4-Way Set Associative Cache Memory

2. At Power on, reset the entire Random Array to all 0's.

3. Random array remains untouched (i.e doesn't get updated) in the case of a Cache Hit.

4. Random array remains untouched in the case of a Snoop Invalidate.

Table 7: Full Random Algorithm as Applied to a 4-Way Set Associative Cache Memory

---

5. Use the following algorithm to replace an entry and update the Random array in the case of a Cache Miss.

```
if(random[1:0] for the indexed entry is 2'b00)
    Replace the indexed entry in Way0
    Set random[1:0] with the value from RNGL while you are updating the entry in Way0
elseif(random[1:0] for the indexed entry is 2'b01)
    Replace the indexed entry in Way1
    Set random[1:0] with the value from RNGL while you are updating the entry in Way1
elseif(random[1:0] for the indexed entry is 2'b10)
    Replace the indexed entry in Way2
    Set random[1:0] with the value from RNGL while you are updating the entry in Way2
else(random[1:0] for the indexed entry is 2'b11)
    Replace the indexed entry in Way3
    Set random[1:0] with the value from RNGL while you are updating the entry in Way3
```

---

13. Describe the various flavours of Pseudo LRU Algorithms that gets used in replacing an entry in a Fully Associative TLB and a 4-Way Set Associative Cache Memory?

Table below shows the various flavours of Pseudo LRU Algorithms that gets used in replacing an entry in a Fully Associative and a 4-Way Set Associative Cache Memory.

Table 8: Pseudo LRU Algorithms

1	3-bit UVL Pseudo LRU algorithm as applied to a Fully Associative TLB
	1. Maintain a 3-bit UVL vector (U (**Used**), V (**Valid**) and L (**Lock**)) for each entry in the Fully Associative TLB as shown in Figure 26 below. 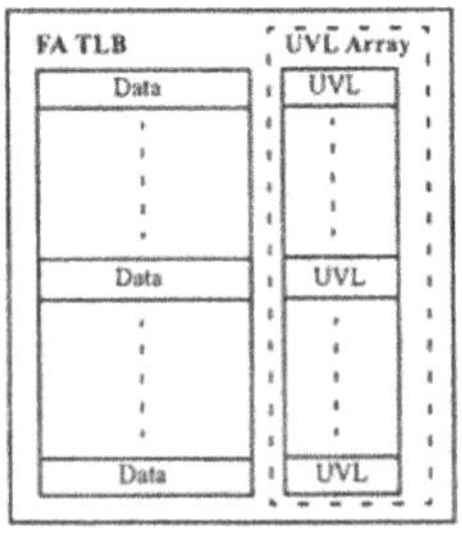 Figure 26: 3-bit UVL Pseudo LRU for a Fully Associative TLB 2. At Power on, reset the entire UVL Array to all 0's. 3. Use the following algorithm to update the 3-bit UVL vector in the case of a TLB Hit `if(TLB access results in a Hit)` `Set the Used bit for the entry which had a Hit to Logic1` `Valid and Lock bits for that entry remain unchanged`

Table 8: Pseudo LRU Algorithms

	4. Use the following algorithm to replace an entry and update the 3-bit UVL vector in the case of a TLB Miss
	
	if(Invalid entry exists in **TLB)**
	   Replace the first **Invalid** entry
	   Set the **Valid** and **Used** bit for the replaced entry to Logic1 when the entry gets updated with the Missed data
	   Set the **Lock** bit for the replaced entry accordingly depending on if you want to lock the Missed data
	elseif(Unused and **Unlocked** entry exists in **TLB)**
	   Replace the first **Unused** and **Unlocked** entry
	   Set the **Valid** and **Used** bit for the replaced entry to Logic1 when the entry gets updated with the Missed data
	   Set the **Lock** bit for the replaced entry accordingly depending on if you want to lock the Missed data
	elseif(Unlocked entry exists in **TLB)**
	   Clear all the **Used** bits and Replace the first **Unlocked** entry
	   Set the **Valid** and **Used** bit for the replaced entry to Logic1 when the entry gets updated with the Missed data
	   Set the **Lock** bit for the replaced entry accordingly depending on if you want to lock the Missed data
	else
	   Replace the last entry in TLB
	   Set the **Valid** and **Used** bit for the replaced entry to Logic1 when the entry gets updated with the Missed data
	   Set the **Lock** bit for the replaced entry accordingly depending on if you want to lock the Missed data

2	*3-bit Pseudo LRU algorithm as applied to a 4- Way Set Associative Cache Memory*
	1. Maintain a 3-bit LRU Array with number of entries equal to the number of entries in the Tag Array as shown in Figure 27 below. Each entry in the LRU Array maintains a 3-bit LRU vector, '**lru[2:0]**', where '**lru[0]**' represents the last recently used (or most recently used) Way among **Way0** and **Way1** (i.e if '0' represents **Way0** to be the last recently used Way and if '1' represents **Way1**). '**lru[2]**' represents the last recently used Way among **Way2** and **Way3** (i.e if '0' represents **Way2** to be last recently used Way and if '1' represents **Way3**) and '**lru[1]**' represents the last recently used Ways among **Way0/Way1** and **Way2/Way3** (i.e if '0' represents **Way0/Way1** to be the last recently used Ways and if '1' represents **Way2/Way3**). In figure below the bits used to index the Data and Tag array may be same or different (i.e index0 = index1 or index0 != index1). 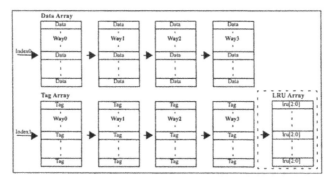 Figure 27: 3-bit Pseudo LRU for a 4-Way Set Associative Cache Memory 2. At Power on, reset the entire LRU Array to all 0's.

Table 8: Pseudo LRU Algorithms

3. Use the following algorithm to update the 3-bit LRU vector in the case of a Cache Hit

```
if(Cache access Hits in Way0)
    Set lru[1:0] for the entry in Way0 which had a Hit to 2'b00
    lru[2] for that entry in Way0 remains unchanged
elseif(Cache access Hits in Way1)
    Set lru[1:0] for the entry in Way1 which had a Hit to 2'b01
    lru[2] for that entry in Way1 remains unchanged
elseif(Cache access Hits in Way2)
    Set lru[2:1] for the entry in Way2 which had a Hit to 2'b01
    lru[0] for that entry in Way2 remains unchanged
else(Cache access Hits in Way3)
    Set lru[2:1] for the entry in Way3 which had a Hit to 2'b11
    lru[0] for that entry in Way3 remains unchanged
```

4. Use the following algorithm to update the 3-bit LRU vector in the case of a Snoop Invalidate

```
if(Snoop Invalidate Hits in Way0)
    Set lru[1:0] for the entry in Way0 which had a Hit to 2'b11
    lru[2] for that entry in Way0 remains unchanged
elseif(Snoop Invalidate Hits in Way1)
    Set lru[1:0] for the entry in Way1 which had a Hit to 2'b10
    lru[2] for that entry in Way1 remains unchanged
elseif(Snoop Invalidate Hits in Way2)
    Set lru[2:1] for the entry in Way2 which had a Hit to 2'b10
    lru[0] for that entry in Way2 remains unchanged
else(Snoop Invalidate Hits in Way3)
    Set lru[2:1] for the entry in Way3 which had a Hit to 2'b00
    lru[0] for that entry in Way3 remains unchanged
```

5. Use the following algorithm to replace an entry and update the 3-bit LRU vector in the case of a Cache Miss

```
if(lru[1]=1)
    if(lru[0]=1)
        Replace the indexed entry in Way0
        Set lru[1:0] for the indexed entry in Way0 to 2'b00 when the entry gets updated with the Missed data
        lru[2] for the indexed entry in Way0 remains unchanged
    else
        Replace the indexed entry in Way1
        Set lru[1:0] for the indexed entry in Way1 to 2'b01 when the entry gets updated with the Missed data
        lru[2] for the indexed entry in Way1 remains unchanged
else
    if(lru[2]=1)
        Replace the indexed entry in Way2
        Set lru[2:1] for the indexed entry in Way2 to 2'b01 when the entry gets updated with the Missed data
        lru[0] for the indexed entry in Way2 remains unchanged
    else
        Replace the indexed entry in Way3
        Set lru[2:1] for the indexed entry in Way3 to 2'b11 when the entry gets updated with the Missed data
        lru[0] for the indexed entry in Way3 remains unchanged
```
 |

| 3 | 3-bit UVL Pseudo LRU algorithm as applied to a 4-Way Set Associative Cache Memory |

Table 8: Pseudo LRU Algorithms

1. Maintain a 3-bit UVL vector (U (**Used**), V (**Valid**) and L (**Lock**)) for each Tag entry in each Way of the 4-Way Set Associative Cache Memory as shown in Figure 28 below. In figure below the bits used to index the Data and Tag array may be same or different (i.e index0 = index1 or index0 != index1).

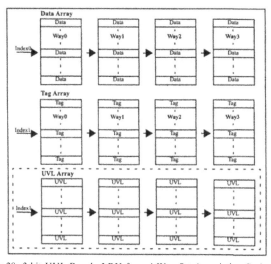

Figure 28: 3-bit UVL Pseudo LRU for a 4-Way Set Associative Cache Memory

2. At Power on, reset UVL arrays corresponding to all 4 Ways to all 0's.

3. Use the following algorithm to update the 3-bit UVL vector in the case of a Cache Hit

```
if(Cache access Hits in Way0)
    Set the Used bit for the entry in Way0 which had a Hit to Logic1
    Valid and Lock bits for that entry in Way0 remain unchanged
elseif(Cache access Hits in Way1)
    Set the Used bit for the entry in Way1 which had a Hit to Logic1
    Valid and Lock bits for that entry in Way1 remain unchanged
elseif(Cache access Hits in Way2)
    Set the Used bit for the entry in Way2 which had a Hit to Logic1
    Valid and Lock bits for that entry in Way2 remain unchanged
elseif(Cache access Hits in Way3)
    Set the Used bit for the entry in Way3 which had a Hit to Logic1
    Valid and Lock bits for that entry in Way3 remain unchanged
```

4. Use the following algorithm to update the 3-bit UVL vector in the case of a Snoop Invalidate

```
if(Snoop Invalidate Hits in Way0)
    Set the Valid bit for the entry in Way0 which had a Hit to Logic0
    Used and Lock bits for that entry in Way0 remain unchanged
elseif(Snoop Invalidate Hits in Way1)
    Set the Valid bit for the entry in Way1 which had a Hit to Logic0
    Used and Lock bits for that entry in Way1 remain unchanged
elseif(Snoop Invalidate Hits in Way2)
    Set the Valid bit for the entry in Way2 which had a Hit to Logic0
    Used and Lock bits for that entry in Way2 remain unchanged
elseif(Snoop Invalidate Hits in Way3)
    Set the Valid bit for the entry in Way3 which had a Hit to Logic0
    Used and Lock bits for that entry in Way3 remain unchanged
```

Table 8: Pseudo LRU Algorithms

	5. Use the following algorithm to replace an entry and update the 3-bit UVL vector in the case of a Cache Miss
	
	If(Indexed entry in **Way0** is **Invalid**)
	    Replace the indexed entry in **Way0**
	    Set the **Valid** and **Used** bit for the indexed entry in **Way0** to Logic1 when the entry gets updated with the Missed data
	    Set the **Lock** bit for the indexed entry in **Way0** accordingly depending on if you want to lock the Missed data
	elseif(Indexed entry in **Way1** is **Invalid**)
	    Replace the indexed entry in **Way1**
	    Set the **Valid** and **Used** bit for the indexed entry in **Way1** to Logic1 when the entry gets updated with the Missed data
	    Set the **Lock** bit for the indexed entry in **Way1** accordingly depending on if you want to lock the Missed data
	elseif(Indexed entry in **Way2** is **Invalid**)
	    Replace the indexed entry in **Way2**
	    Set the **Valid** and **Used** bit for the indexed entry in **Way2** to Logic1 when the entry gets updated with the Missed data
	    Set the **Lock** bit for the indexed entry in **Way2** accordingly depending on if you want to lock the Missed data
	elseif(Indexed entry in **Way3** is **Invalid**)
	    Replace the indexed entry in **Way3**
	    Set the **Valid** and **Used** bit for the indexed entry in **Way3** to Logic1 when the entry gets updated with the Missed data
	    Set the **Lock** bit for the indexed entry in **Way3** accordingly depending on if you want to lock the Missed data
	elseif(Indexed entry in **Way0** is **Unused** and **Unlocked**)
	    Replace the indexed entry in **Way0**
	    Set the **Valid** and **Used** bit for the indexed entry in **Way0** to Logic1 when the entry gets updated with the Missed data
	    Set the **Lock** bit for the indexed entry in **Way0** accordingly depending on if you want to lock the Missed data
	elseif(Indexed entry in **Way1** is **Unused** and **Unlocked**)
	    Replace the indexed entry in **Way1**
	    Set the **Valid** and **Used** bit for the indexed entry in **Way1** to Logic1 when the entry gets updated with the Missed data
	    Set the **Lock** bit for the indexed entry in **Way1** accordingly depending on if you want to lock the Missed data
	elseif(Indexed entry in **Way2** is **Unused** and **Unlocked**)
	    Replace the indexed entry in **Way2**
	    Set the **Valid** and **Used** bit for the indexed entry in **Way2** to Logic1 when the entry gets updated with the Missed data
	    Set the **Lock** bit for the indexed entry in **Way2** accordingly depending on if you want to lock the Missed data
	elseif(Indexed entry in **Way3** is **Unused** and **Unlocked**)
	    Replace the indexed entry in **Way3**
	    Set the **Valid** and **Used** bit for the indexed entry in **Way3** to Logic1 when the entry gets updated with the Missed data
	    Set the **Lock** bit for the indexed entry in **Way3** accordingly depending on if you want to lock the Missed data
	elseif(Indexed entry in **Way0** is **Unlocked**)
	    Clear the **Used** bits for the indexed entries in all 4 **Ways** and Replace the indexed entry in **Way0**
	    Set the **Valid** and **Used** bit for the indexed entry in **Way0** to Logic1 when the entry gets updated with the Missed data
	    Set the **Lock** bit for the indexed entry in **Way0** accordingly depending on if you want to lock the Missed data
	elseif(Indexed entry in **Way1** is **Unlocked**)
	    Clear the **Used** bits for the indexed entries in all 4 **Ways** and Replace the indexed entry in **Way1**
	    Set the **Valid** and **Used** bit for the indexed entry in **Way1** to Logic1 when the entry gets updated with the Missed data
	    Set the **Lock** bit for the indexed entry in **Way1** accordingly depending on if you want to lock the Missed data
	elseif(Indexed entry in **Way2** is **Unlocked**)
	    Clear the **Used** bits for the indexed entries in all 4 **Ways** and Replace the indexed entry in **Way2**
	    Set the **Valid** and **Used** bit for the indexed entry in **Way2** to Logic1 when the entry gets updated with the Missed data
	    Set the **Lock** bit for the indexed entry in **Way2** accordingly depending on if you want to lock the Missed data
	elseif(Indexed entry in **Way3** is **Unlocked**)
	    Clear the **Used** bits for the indexed entries in all 4 **Ways** and Replace the indexed entry in **Way3**
	    Set the **Valid** and **Used** bit for the indexed entry in **Way3** to Logic1 when the entry gets updated with the Missed data
	    Set the **Lock** bit for the indexed entry in **Way3** accordingly depending on if you want to lock the Missed data
	else
	    Replace the indexed entry in **Way0**
	    Set the **Valid** and **Used** bit for the indexed entry in **Way0** to Logic1 when the entry gets updated with the Missed data
	    Set the **Lock** bit for the indexed entry in **Way0** accordingly depending on if you want to lock the Missed data

4	*3-bit UVL-RR Pseudo LRU algorithm as applied to a 4-Way Set Associative Cache Memory*
	1. Maintain a 3-bit UVL vector (U (**Used**), V (**Valid**) and L (**Lock**)) for each Tag entry in each Way of the 4-Way Set Associative Cache Memory as shown in Figure 29 below. Also maintain a single round robin bit (**round_robin**) for the entire 4-Way Set Associative Cache Memory as shown in figure below. In figure below the bits used to index the Data and Tag array may be same or different (i.e index0 = index1 or index0 != index1).

Table 8: Pseudo LRU Algorithms

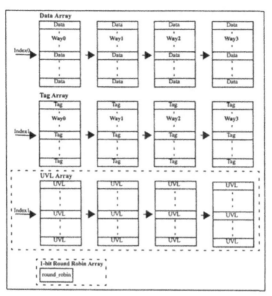

Figure 29: 3-bit UVL-RR Pseudo LRU for a 4-Way Set Associative Cache Memory

2. At Power on, reset UVL arrays corresponding to all 4 Ways to all 0's including the round robin bit.

3. Use the following algorithm to update the 3-bit UVL vector in the case of a Cache Hit

```
if(Cache access Hits in Way0)
    Set the Used bit for the entry in Way0 which had a Hit to Logic1
    Valid and Lock bits for that entry in Way0 remain unchanged
elseif(Cache access Hits in Way1)
    Set the Used bit for the entry in Way1 which had a Hit to Logic1
    Valid and Lock bits for that entry in Way1 remain unchanged
elseif(Cache access Hits in Way2)
    Set the Used bit for the entry in Way2 which had a Hit to Logic1
    Valid and Lock bits for that entry in Way2 remain unchanged
elseif(Cache access Hits in Way3)
    Set the Used bit for the entry in Way3 which had a Hit to Logic1
    Valid and Lock bits for that entry in Way3 remain unchanged
```

4. Use the following algorithm to update the 3-bit UVL vector in the case of a Snoop Invalidate

```
if(Snoop Invalidate Hits in Way0)
    Set the Valid bit for the entry in Way0 which had a Hit to Logic0
    Used and Lock bits for that entry in Way0 remain unchanged
elseif(Snoop Invalidate Hits in Way1)
    Set the Valid bit for the entry in Way1 which had a Hit to Logic0
    Used and Lock bits for that entry in Way1 remain unchanged
elseif(Snoop Invalidate Hits in Way2)
    Set the Valid bit for the entry in Way2 which had a Hit to Logic0
    Used and Lock bits for that entry in Way2 remain unchanged
elseif(Snoop Invalidate Hits in Way3)
    Set the Valid bit for the entry in Way3 which had a Hit to Logic0
    Used and Lock bits for that entry in Way3 remain unchanged
```

Table 8: Pseudo LRU Algorithms

	5. Use the following algorithm to replace an entry and update the 3-bit UVL vector and round_robin bit in the case of a Cache Miss

	if(round_robin=1'b0)
	  if(Indexed entry in **Way0** is **Invalid**)
	    Replace the indexed entry in **Way0**
	    Set the **Valid** and **Used** bit for the indexed entry in **Way0** to Logic1 when the entry gets updated with the Missed data
	    Set the **Lock** bit for the indexed entry in **Way0** accordingly depending on if you want to lock the Missed data
	  elseif(Indexed entry in **Way1** is **Invalid**)
	    Replace the indexed entry in **Way1**
	    Set the **Valid** and **Used** bit for the indexed entry in **Way1** to Logic1 when the entry gets updated with the Missed data
	    Set the **Lock** bit for the indexed entry in **Way1** accordingly depending on if you want to lock the Missed data
	  elseif(Indexed entry in **Way0** is **Unused** and **Unlocked**)
	    Replace the indexed entry in **Way0**
	    Set the **Valid** and **Used** bit for the indexed entry in **Way0** to Logic1 when the entry gets updated with the Missed data
	    Set the **Lock** bit for the indexed entry in **Way0** accordingly depending on if you want to lock the Missed data
	  elseif(Indexed entry in **Way1** is **Unused** and **Unlocked**)
	    Replace the indexed entry in **Way1**
	    Set the **Valid** and **Used** bit for the indexed entry in **Way1** to Logic1 when the entry gets updated with the Missed data
	    Set the **Lock** bit for the indexed entry in **Way1** accordingly depending on if you want to lock the Missed data
	  elseif(Indexed entry in **Way0** is **Unlocked**)
	    Clear the **Used** bits for the indexed entries in **Way0** and **Way1** and Replace the indexed entry in **Way0**
	    Set the **Valid** and **Used** bit for the indexed entry in **Way0** to Logic1 when the entry gets updated with the Missed data
	    Set the **Lock** bit for the indexed entry in **Way0** accordingly depending on if you want to lock the Missed data
	  elseif(Indexed entry in **Way1** is **Unlocked**)
	    Clear the **Used** bits for the indexed entries in **Way0** and **Way1** and Replace the indexed entry in **Way1**
	    Set the **Valid** and **Used** bit for the indexed entry in **Way1** to Logic1 when the entry gets updated with the Missed data
	    Set the **Lock** bit for the indexed entry in **Way1** accordingly depending on if you want to lock the Missed data
	  else
	    Replace the indexed entry in **Way0**
	    Set the **Valid** and **Used** bit for the indexed entry in **Way0** to Logic1 when the entry gets updated with the Missed data
	    Set the **Lock** bit for the indexed entry in **Way0** accordingly depending on if you want to lock the Missed data
	    Set **round_robin** bit to Logic1
	else
	  if(Indexed entry in **Way2** is **Invalid**)
	    Replace the indexed entry in **Way2**
	    Set the **Valid** and **Used** bit for the indexed entry in **Way2** to Logic1 when the entry gets updated with the Missed data
	    Set the **Lock** bit for the indexed entry in **Way2** accordingly depending on if you want to lock the Missed data
	  elseif(Indexed entry in **Way3** is **Invalid**)
	    Replace the indexed entry in **Way3**
	    Set the **Valid** and **Used** bit for the indexed entry in **Way3** to Logic1 when the entry gets updated with the Missed data
	    Set the **Lock** bit for the indexed entry in **Way3** accordingly depending on if you want to lock the Missed data
	  elseif(Indexed entry in **Way2** is **Unused** and **Unlocked**)
	    Replace the indexed entry in **Way2**
	    Set the **Valid** and **Used** bit for the indexed entry in **Way2** to Logic1 when the entry gets updated with the Missed data
	    Set the **Lock** bit for the indexed entry in **Way2** accordingly depending on if you want to lock the Missed data
	  elseif(Indexed entry in **Way3** is **Unused** and **Unlocked**)
	    Replace the indexed entry in **Way3**
	    Set the **Valid** and **Used** bit for the indexed entry in **Way3** to Logic1 when the entry gets updated with the Missed data
	    Set the **Lock** bit for the indexed entry in **Way3** accordingly depending on if you want to lock the Missed data
	  elseif(Indexed entry in **Way2** is **Unlocked**)
	    Clear the **Used** bits for the indexed entries in **Way2** and **Way3** and Replace the indexed entry in **Way2**
	    Set the **Valid** and **Used** bit for the indexed entry in **Way2** to Logic1 when the entry gets updated with the Missed data
	    Set the **Lock** bit for the indexed entry in **Way2** accordingly depending on if you want to lock the Missed data
	  elseif(Indexed entry in **Way3** is **Unlocked**)
	    Clear the **Used** bits for the indexed entries in **Way2** and **Way3** and Replace the indexed entry in **Way3**
	    Set the **Valid** and **Used** bit for the indexed entry in **Way3** to Logic1 when the entry gets updated with the Missed data
	    Set the **Lock** bit for the indexed entry in **Way3** accordingly depending on if you want to lock the Missed data
	  else
	    Replace the indexed entry in **Way2**
	    Set the **Valid** and **Used** bit for the indexed entry in **Way2** to Logic1 when the entry gets updated with the Missed data
	    Set the **Lock** bit for the indexed entry in **Way2** accordingly depending on if you want to lock the Missed data
	    Set **round_robin** bit to Logic0

5	*8-bit Pseudo LRU algorithm as applied to a 4-Way Set Associative Cache Memory*

Table 8: Pseudo LRU Algorithms

1. Maintain a 8-bit LRU Array with number of entries equal to the number of entries in the Tag Array as shown in Figure 30 below. Each entry in the LRU Array maintains a 8-bit LRU vector, 'lru[7:0]', where 'lru[1:0]' represents the status of **Way0** (i.e if 2'b00 represents **Way0** to be the least recently used Way among all 4 Ways, if 2'01 represents **Way0** to be the second least recently used Way among all 4 Ways, if 2'b10 represents **Way0** to be the third least recently used Way among all 4 Ways and if 2'b11 represents **Way0** to be the last recently used (or most recently used) Way among all 4 Ways), 'lru[3:2]' represents the status of **Way1**, 'lru[5:4]' represents the status of **Way2** and 'lru[7:6]' represents the status of **Way3**. In figure below the bits used to index the Data and Tag array may be same or different (i.e index0 = index1 or index0 != index1).

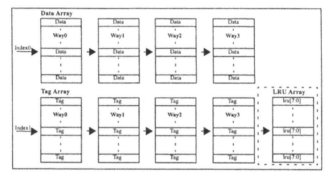

Figure 30: 8-bit Pseudo LRU for a 4-Way Set Associative Cache Memory

2. At Power on, set each entry in LRU Array to 8'b11100100 (i.e **lru[7:0]** = 8'b11100100 for all entries).

3. Use the following algorithm to update the 8-bit vector in the case of a Cache Hit

```
If(Cache access Hits in Way0)
    if(lru[1:0]=2'b00) lru[7:6]=(lru[7:6]-1), lru[5:4]=(lru[5:4]-1), lru[3:2]=(lru[3:2]-1), lru[1:0]=2'b11
    elseif(lru[1:0]=2'b01)
        if(lru[3:2]=2'b00) lru[3:2]=lru[3:2] else lru[3:2]=(lru[3:2]-1)
        if(lru[5:4]=2'b00) lru[5:4]=lru[5:4] else lru[5:4]=(lru[5:4]-1)
        if(lru[7:6]=2'b00) lru[7:6]=lru[7:6] else lru[7:6]=(lru[7:6]-1)
        lru[1:0]=2'b11
        elseif(lru[1:0]=2'b10)
            if(lru[3:2]=2'b11) lru[3:2]=(lru[3:2]-1) else lru[3:2]=lru[3:2]
            if(lru[5:4]=2'b11) lru[5:4]=(lru[5:4]-1) else lru[5:4]=lru[5:4]
            if(lru[7:6]=2'b11) lru[7:6]=(lru[7:6]-1) else lru[7:6]=lru[7:6]
            lru[1:0]=2'b11
        else lru[7:6]=lru[7:6], lru[5:4]=lru[5:4], lru[3:2]=lru[3:2], lru[1:0]=lru[1:0]
elseif(Cache access Hits in Way1)
    if(lru[3:2]=2'b00) lru[7:6]=(lru[7:6]-1), lru[5:4]=(lru[5:4]-1), lru[3:2]=2'b11, lru[1:0]=(lru[1:0]-1)
    elseif(lru[3:2]=2'b01)
        if(lru[1:0]=2'b00) lru[1:0]=lru[1:0] else lru[1:0]=(lru[1:0]-1)
        if(lru[5:4]=2'b00) lru[5:4]=lru[5:4] else lru[5:4]=(lru[5:4]-1)
        if(lru[7:6]=2'b00) lru[7:6]=lru[7:6] else lru[7:6]=(lru[7:6]-1)
        lru[3:2]=2'b11
        elseif(lru[3:2]=2'b10)
            if(lru[1:0]=2'b11) lru[1:0]=(lru[1:0]-1) else lru[1:0]=lru[1:0]
            if(lru[5:4]=2'b11) lru[5:4]=(lru[5:4]-1) else lru[5:4]=lru[5:4]
            if(lru[7:6]=2'b11) lru[7:6]=(lru[7:6]-1) else lru[7:6]=lru[7:6]
            lru[3:2]=2'b11
        else lru[7:6]=lru[7:6], lru[5:4]=lru[5:4], lru[3:2]=lru[3:2], lru[1:0]=lru[1:0]
```

Table 8: Pseudo LRU Algorithms

```
elseif(Cache access Hits in Way2)
    if(lru[5:4]=2'b00) lru[7:6]=(lru[7:6]-1), lru[5:4]=2'b11, lru[3:2]=(lru[3:2]-1), lru[1:0]=(lru[1:0]-1)
    elseif(lru[5:4]=2'b01)
            if(lru[1:0]=2'b00) lru[1:0]=lru[1:0] else lru[1:0]=(lru[1:0]-1)
            if(lru[3:2]=2'b00) lru[3:2]=lru[3:2] else lru[3:2]=(lru[3:2]-1)
            if(lru[7:6]=2'b00) lru[7:6]=lru[7:6] else lru[7:6]=(lru[7:6]-1)
            lru[5:4]=2'b11
        elseif(lru[5:4]=2'b10)
                if(lru[1:0]=2'b11) lru[1:0]=(lru[1:0]-1) else lru[1:0]=lru[1:0]
                if(lru[3:2]=2'b11) lru[3:2]=(lru[3:2]-1) else lru[3:2]=lru[3:2]
                if(lru[7:6]=2'b11) lru[7:6]=(lru[7:6]-1) else lru[7:6]=lru[7:6]
                lru[5:4]=2'b11
            else lru[5:4]=lru[7:6], lru[5:4]=lru[5:4], lru[3:2]=lru[3:2], lru[1:0]=lru[1:0]
    elseif(Cache access Hits in Way3)
        if(lru[7:6]=2'b00) lru[7:6]=2'b11, lru[5:4]=(lru[5:4]-1), lru[3:2]=(lru[3:2]-1), lru[1:0]=(lru[1:0]-1)
        elseif(lru[7:6]=2'b01)
                if(lru[1:0]=2'b00) lru[1:0]=lru[1:0] else lru[1:0]=(lru[1:0]-1)
                if(lru[3:2]=2'b00) lru[3:2]=lru[3:2] else lru[3:2]=(lru[3:2]-1)
                if(lru[5:4]=2'b00) lru[5:4]=lru[5:4] else lru[5:4]=(lru[5:4]-1)
                lru[7:6]=2'b11
            elseif(lru[7:6]=2'b10)
                    if(lru[1:0]=2'b11) lru[1:0]=(lru[1:0]-1) else lru[1:0]=lru[1:0]
                    if(lru[3:2]=2'b11) lru[3:2]=(lru[3:2]-1) else lru[3:2]=lru[3:2]
                    if(lru[5:4]=2'b11) lru[5:4]=(lru[5:4]-1) else lru[5:4]=lru[5:4]
                    lru[7:6]=2'b11
                else lru[7:6]=lru[7:6], lru[5:4]=lru[5:4], lru[3:2]=lru[3:2], lru[1:0]=lru[1:0]
```

4. Use the following algorithm to update the 8-bit vector in the case of a Snoop Invalidate

```
if(Snoop Invalidate Hits in Way0)
    if(lru[1:0]=2'b11) lru[7:6]=(lru[7:6]+1), lru[5:4]=(lru[5:4]+1), lru[3:2]=(lru[3:2]+1), lru[1:0]=2'b00
    elseif(lru[1:0]=2'b10)
            if(lru[3:2]=2'b11) lru[3:2]=lru[3:2] else lru[3:2]=(lru[3:2]+1)
            if(lru[5:4]=2'b11) lru[5:4]=lru[5:4] else lru[5:4]=(lru[5:4]+1)
            if(lru[7:6]=2'b11) lru[7:6]=lru[7:6] else lru[7:6]=(lru[7:6]+1)
            lru[1:0]=2'b00
        elseif(lru[1:0]=2'b01)
                if(lru[3:2]=2'b00) lru[3:2]=(lru[3:2]+1) else lru[3:2]=lru[3:2]
                if(lru[5:4]=2'b00) lru[5:4]=(lru[5:4]+1) else lru[5:4]=lru[5:4]
                if(lru[7:6]=2'b00) lru[7:6]=(lru[7:6]+1) else lru[7:6]=lru[7:6]
                lru[1:0]=2'b00
            else lru[7:6]=lru[7:6], lru[5:4]=lru[5:4], lru[3:2]=lru[3:2], lru[1:0]=lru[1:0]
    elseif(Snoop Invalidate Hits in Way1)
        if(lru[3:2]=2'b11) lru[7:6]=(lru[7:6]+1), lru[5:4]=(lru[5:4]+1), lru[3:2]=2'b00, lru[1:0]=(lru[1:0]+1)
        elseif(lru[3:2]=2'b10)
                if(lru[1:0]=2'b11) lru[1:0]=lru[1:0] else lru[1:0]=(lru[1:0]+1)
                if(lru[5:4]=2'b11) lru[5:4]=lru[5:4] else lru[5:4]=(lru[5:4]+1)
                if(lru[7:6]=2'b11) lru[7:6]=lru[7:6] else lru[7:6]=(lru[7:6]+1)
                lru[3:2]=2'b00
            elseif(lru[3:2]=2'b01)
                    if(lru[1:0]=2'b00) lru[1:0]=(lru[1:0]+1) else lru[1:0]=lru[1:0]
                    if(lru[5:4]=2'b00) lru[5:4]=(lru[5:4]+1) else lru[5:4]=lru[5:4]
                    if(lru[7:6]=2'b00) lru[7:6]=(lru[7:6]+1) else lru[7:6]=lru[7:6]
                    lru[3:2]=2'b00
                else lru[7:6]=lru[7:6], lru[5:4]=lru[5:4], lru[3:2]=lru[3:2], lru[1:0]=lru[1:0]
    elseif(Snoop Invalidate Hits in Way2)
        if(lru[5:4]=2'b11) lru[7:6]=(lru[7:6]+1), lru[5:4]=2'b00, lru[3:2]=(lru[3:2]+1), lru[1:0]=(lru[1:0]+1)
        elseif(lru[5:4]=2'b10)
                if(lru[1:0]=2'b11) lru[1:0]=lru[1:0] else lru[1:0]=(lru[1:0]+1)
                if(lru[3:2]=2'b11) lru[3:2]=lru[3:2] else lru[3:2]=(lru[3:2]+1)
                if(lru[7:6]=2'b11) lru[7:6]=lru[7:6] else lru[7:6]=(lru[7:6]+1)
                lru[5:4]=2'b00
            elseif(lru[5:4]=2'b01)
                    if(lru[1:0]=2'b00) lru[1:0]=(lru[1:0]+1) else lru[1:0]=lru[1:0]
                    if(lru[3:2]=2'b00) lru[3:2]=(lru[3:2]+1) else lru[3:2]=lru[3:2]
                    if(lru[7:6]=2'b00) lru[7:6]=(lru[7:6]+1) else lru[7:6]=lru[7:6]
                    lru[5:4]=2'b00
                else lru[5:4]=lru[7:6], lru[5:4]=lru[5:4], lru[3:2]=lru[3:2], lru[1:0]=lru[1:0]
```

Table 8: Pseudo LRU Algorithms

```
elseif(Snoop Invalidate Hits in Way3)
    if(lru[7:6]=2'b11) lru[7:6]=2'b00, lru[5:4]=(lru[5:4]+1), lru[3:2]=(lru[3:2]+1), lru[1:0]=(lru[1:0]+1)
    elseif(lru[7:6]=2'b10)
        if(lru[1:0]=2'b11) lru[1:0]=lru[1:0] else lru[1:0]=(lru[1:0]+1)
        if(lru[3:2]=2'b11) lru[3:2]=lru[3:2] else lru[3:2]=(lru[3:2]+1)
        if(lru[5:4]=2'b11) lru[5:4]=lru[5:4] else lru[5:4]=(lru[5:4]+1)
        lru[7:6]=2'b00
    elseif(lru[7:6]=2'b01)
        if(lru[1:0]=2'b00) lru[1:0]=(lru[1:0]+1) else lru[1:0]=lru[1:0]
        if(lru[3:2]=2'b00) lru[3:2]=(lru[3:2]+1) else lru[3:2]=lru[3:2]
        if(lru[5:4]=2'b00) lru[5:4]=(lru[5:4]+1) else lru[5:4]=lru[5:4]
        lru[7:6]=2'b00
    else lru[7:6]=lru[7:6], lru[5:4]=lru[5:4], lru[3:2]=lru[3:2], lru[1:0]=lru[1:0]
```

5. Use the following algorithm to replace an entry and update the 8-bit LRU vector in the case of a Cache Miss

```
if(lru|1:0|=2'b00)
    Replace entry in Way0
    Update the LRU vector corresponding to the entry with the following values while it gets updated with the Missed data
    lru[7:6]=(lru[7:6]-1), lru[5:4]=(lru[5:4]-1), lru[3:2]=(lru[3:2]-1), lru[1:0]=2'b11
elseif(lru[3:2]=2'b00)
    Replace entry in Way1
    Update the LRU vector corresponding to the entry with the following values while it gets updated with the Missed data
    lru[7:6]=(lru[7:6]-1), lru[5:4]=(lru[5:4]-1), lru[3:2]=2'b11, lru[1:0]=(lru[1:0]-1)
elseif(lru[5:4]=2'b00)
    Replace entry in Way2
    Update the LRU vector corresponding to the entry with the following values while it gets updated with the Missed data
    lru[7:6]=(lru[7:6]-1), lru[5:4]=2'b11, lru[3:2]=(lru[3:2]-1), lru[1:0]=(lru[1:0]-1)
else
    Replace entry in Way3
    Update the LRU vector corresponding to the entry with the following values while it gets updated with the Missed data
    lru[7:6]=2'b11, lru[5:4]=(lru[5:4]-1), lru[3:2]=(lru[3:2]-1), lru[1:0]=(lru[1:0]-1)
```

14. Describe the way Full LRU Algorithm gets used in replacing an entry in a 4-Way Set Associative Cache Memory?

Table below shows the way Full LRU Algorithm gets used in replacing an entry in a 4-Way Set Associative Cache Memory.

Table 9: Full LRU Algorithm as Applied to a 4-Way Set Associative Cache Memory

1. Maintain a N-bit LRU vector for each Tag entry in each Way (i.e **lru0[(N-1):0]** for Way0, **lru1[(N-1):0]** for Way1, **lru2[(N-1):0]** for Way2 and **lru3[(N-1):0]** for Way3) of the 4-Way Set Associative Cache Memory as shown in Figure 31 below. Also maintain a N-bit (i.e **lru[(N-1):0]**) Counter Array with number of entries equal to the number of entries in the Tag Array as shown in figure below. In figure below the bits used to index the Data and Tag array may be same or different (i.e index0 = index 1 or index0 != index1).

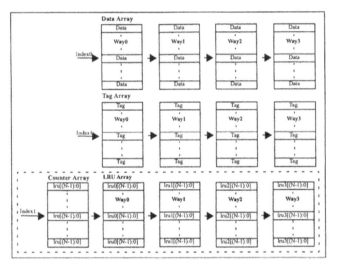

Figure 31: Full LRU for a 4-Way Set Associative Cache Memory

2. At Power on, reset the N-bit Counter Array to all decimal 3's, N-bit LRU Array corresponding to **Way0** to all decimal 0's, N-bit LRU Array corresponding to **Way1** to all decimal 1's, N-bit LRU Array corresponding to **Way2** to all decimal 2's and N-bit LRU Array corresponding to **Way3** to all decimal 3's.

Table 9: Full LRU Algorithm as Applied to a 4-Way Set Associative Cache Memory

3. Use the following algorithm to update the N-bit Counter and N-bit LRU vectors in the case of a Cache Hit

```
if(Cache access Hits in Way0)
    if(N-bit counter corresponding to the entry which had a HIT is at its max count)
        Set N-bit counter and lru0[(N-1):0] to decimal 3
        if(lru1[(N-1):0] corresponding to the entry is greater than lru2[(N-1):0] and lru3[(N-1):0])
            Set lru1[(N-1):0] to decimal 2
            if(lru2[(N-1):0] corresponding to the entry is greater than lru3[(N-1):0])
                Set lru2[(N-1):0] to decimal 1 and lru3[(N-1):0] to decimal 0
            else Set lru3[(N-1):0] to decimal 1 and lru2[(N-1):0] to decimal 0
        elseif(lru2[(N-1):0] corresponding to the entry is greater than lru1[(N-1):0] and lru3[(N-1):0])
            Set lru2[(N-1):0] to decimal 2
            if(lru1[(N-1):0] corresponding to the entry is greater than lru3[(N-1):0])
                Set lru1[(N-1):0] to decimal 1 and lru3[(N-1):0] to decimal 0
            else Set lru3[(N-1):0] to decimal 1 and lru1[(N-1):0] to decimal 0
        else Set lru3[(N-1):0] corresponding to the entry to decimal 2
            if(lru1[(N-1):0] corresponding to the entry is greater than lru2[(N-1):0])
                Set lru1[(N-1):0] to decimal 1 and lru2[(N-1):0] to decimal 0
            else Set lru2[(N-1):0] to decimal 1 and lru1[(N-1):0] to decimal 0
    else
        Increment the N-bit Counter corresponding to the entry by 1
        Set lru vector(i.e lru0[(N-1):0]) for the entry in Way0 which had a Hit with the incremented value of N-bit Counter
        corresponding to the entry
        lru vectors corresponding to Way1, Way2 and Way3 i.e lru1[(N-1):0], lru2[(N-1):0] and lru3[(N-1):0] remain unchanged
elseif(Cache access Hits in Way1)
    if(N-bit counter corresponding to the entry which had a HIT is at its max count)
        Set N-bit counter and lru1[(N-1):0] to decimal 3
        if(lru0[(N-1):0] corresponding to the entry is greater than lru2[(N-1):0] and lru3[(N-1):0])
            Set lru0[(N-1):0] to decimal 2
            if(lru2[(N-1):0] corresponding to the entry is greater than lru3[(N-1):0])
                Set lru2[(N-1):0] to decimal 1 and lru3[(N-1):0] to decimal 0
            else Set lru3[(N-1):0] to decimal 1 and lru2[(N-1):0] to decimal 0
        elseif(lru2[(N-1):0] corresponding to the entry is greater than lru0[(N-1):0] and lru3[(N-1):0])
            Set lru2[(N-1):0] to decimal 2
            if(lru0[(N-1):0] corresponding to the entry is greater than lru3[(N-1):0])
                Set lru0[(N-1):0] to decimal 1 and lru3[(N-1):0] to decimal 0
            else Set lru3[(N-1):0] to decimal 1 and lru0[(N-1):0] to decimal 0
        else Set lru3[(N-1):0] corresponding to the entry to decimal 2
            if(lru0[(N-1):0] corresponding to the entry is greater than lru2[(N-1):0])
                Set lru0[(N-1):0] to decimal 1 and lru2[(N-1):0] to decimal 0
            else Set lru2[(N-1):0] to decimal 1 and lru0[(N-1):0] to decimal 0
    else
        Increment the N-bit Counter corresponding to the entry by 1
        Set lru vector(i.e lru1[(N-1):0]) for the entry in Way1 which had a Hit with the incremented value of N-bit Counter
        corresponding to the entry
        lru vectors corresponding to Way0, Way2 and Way3 i.e lru0[(N-1):0], lru2[(N-1):0] and lru3[(N-1):0] remain unchanged
elseif(Cache access Hits in Way2)
    if(N-bit counter corresponding to the entry which had a HIT is at its max count)
        Set N-bit counter and lru2[(N-1):0] to decimal 3
        if(lru0[(N-1):0] corresponding to the entry is greater than lru1[(N-1):0] and lru3[(N-1):0])
            Set lru0[(N-1):0] to decimal 2
            if(lru1[(N-1):0] corresponding to the entry is greater than lru3[(N-1):0])
                Set lru1[(N-1):0] to decimal 1 and lru3[(N-1):0] to decimal 0
            else Set lru3[(N-1):0] to decimal 1 and lru1[(N-1):0] to decimal 0
        elseif(lru1[(N-1):0] corresponding to the entry is greater than lru0[(N-1):0] and lru3[(N-1):0])
            Set lru1[(N-1):0] to decimal 2
            if(lru0[(N-1):0] corresponding to the entry is greater than lru3[(N-1):0])
                Set lru0[(N-1):0] to decimal 1 and lru3[(N-1):0] to decimal 0
            else Set lru3[(N-1):0] to decimal 1 and lru0[(N-1):0] to decimal 0
        else Set lru3[(N-1):0] corresponding to the entry to decimal 2
            if(lru0[(N-1):0] corresponding to the entry is greater than lru1[(N-1):0])
                Set lru0[(N-1):0] to decimal 1 and lru1[(N-1):0] to decimal 0
            else Set lru1[(N-1):0] to decimal 1 and lru0[(N-1):0] to decimal 0
    else
        Increment the N-bit Counter corresponding to the entry by 1
        Set lru vector(i.e lru2[(N-1):0]) for the entry in Way2 which had a Hit with the incremented value of N-bit Counter
        corresponding to the entry
        lru vectors corresponding to Way0, Way1 and Way3 i.e lru0[(N-1):0], lru1[(N-1):0] and lru3[(N-1):0] remain unchanged
```

Table 9: Full LRU Algorithm as Applied to a 4-Way Set Associative Cache Memory

```
elseif(Cache access Hits in Way3)
    if(N-bit counter corresponding to the entry which had a HIT is at its max count)
        Set N-bit counter and lru3[(N-1):0] to decimal 3
        if(lru0[(N-1):0] corresponding to the entry is greater than lru1[(N-1):0] and lru2[(N-1):0])
            Set lru0[(N-1):0] to decimal 2
            if(lru1[(N-1):0] corresponding to the entry is greater than lru2[(N-1):0])
                Set lru1[(N-1):0] to decimal 1 and lru2[(N-1):0] to decimal 0
            else Set lru2[(N-1):0] to decimal 1 and lru1[(N-1):0] to decimal 0
        elseif(lru1[(N-1):0] corresponding to the entry is greater than lru0[(N-1):0] and lru2[(N-1):0])
            Set lru1[(N-1):0] to decimal 2
            if(lru0[(N-1):0] corresponding to the entry is greater than lru2[(N-1):0])
                Set lru0[(N-1):0] to decimal 1 and lru2[(N-1):0] to decimal 0
            else Set lru2[(N-1):0] to decimal 1 and lru0[(N-1):0] to decimal 0
        else Set lru2[(N-1):0] corresponding to the entry to decimal 2
            if(lru0[(N-1):0] corresponding to the entry is greater than lru1[(N-1):0])
                Set lru0[(N-1):0] to decimal 1 and lru1[(N-1):0] to decimal 0
            else Set lru1[(N-1):0] to decimal 1 and lru0[(N-1):0] to decimal 0
    else
        Increment the N-bit Counter corresponding to the entry by 1
        Set lru vector(i.e lru3[(N-1):0]) for the entry in Way3 which had a Hit with the incremented value of N-bit Counter
        corresponding to the entry
        lru vectors corresponding to Way0, Way1 and Way2 i.e lru0[(N-1):0], lru1[(N-1):0] and lru2[(N-1):0] remain unchanged
```

4. Don't update N-bit Counter and N-bit LRU vectors in the case of Snoop Invalidate as this involves lot of additional complexity.

5. Use the following algorithm to replace an entry and update the N-bit Counter and N-bit LRU vectors in the case of a Cache Miss

```
if(lru0[(N-1):0] has the least count compared to other lru vectors corresponding to the indexed entry)
    Replace entry in Way0
    if(N-bit counter corresponding to the entry which had a MISS is at its max count)
        Set N-bit counter and lru0[(N-1):0] to decimal 3 while you are updating the entry in Way0
        if(lru1[(N-1):0] corresponding to the entry is greater than lru2[(N-1):0] and lru3[(N-1):0])
            Set lru1[(N-1):0] to decimal 2
            if(lru2[(N-1):0] corresponding to the entry is greater than lru3[(N-1):0])
                Set lru2[(N-1):0] to decimal 1 and lru3[(N-1):0] to decimal 0
            else Set lru3[(N-1):0] to decimal 1 and lru2[(N-1):0] to decimal 0
        elseif(lru2[(N-1):0] corresponding to the entry is greater than lru1[(N-1):0] and lru3[(N-1):0])
            Set lru2[(N-1):0] to decimal 2
            if(lru1[(N-1):0] corresponding to the entry is greater than lru3[(N-1):0])
                Set lru1[(N-1):0] to decimal 1 and lru3[(N-1):0] to decimal 0
            else Set lru3[(N-1):0] to decimal 1 and lru1[(N-1):0] to decimal 0
        else Set lru3[(N-1):0] corresponding to the entry to decimal 2
            if(lru1[(N-1):0] corresponding to the entry is greater than lru2[(N-1):0])
                Set lru1[(N-1):0] to decimal 1 and lru2[(N-1):0] to decimal 0
            else Set lru2[(N-1):0] to decimal 1 and lru1[(N-1):0] to decimal 0
    else
        Increment the N-bit Counter corresponding to the entry by 1 while you are updating the entry
        Set lru0[(N-1):0] for the entry in Way0 with the incremented value of N-bit Counter
        lru vectors corresponding to Way1, Way2 and Way3 i.e lru1[(N-1):0], lru2[(N-1):0] and lru3[(N-1):0] remain unchanged
elseif(lru1[(N-1):0] has the least count compared to other lru vectors corresponding to the indexed entry)
    Replace entry in Way1
    if(N-bit counter corresponding to the entry which had a MISS is at its max count)
        Set N-bit counter and lru1[(N-1):0] to decimal 3 while you are updating the entry in Way1
        if(lru0[(N-1):0] corresponding to the entry is greater than lru2[(N-1):0] and lru3[(N-1):0])
            Set lru0[(N-1):0] to decimal 2
            if(lru2[(N-1):0] corresponding to the entry is greater than lru3[(N-1):0])
                Set lru2[(N-1):0] to decimal 1 and lru3[(N-1):0] to decimal 0
            else Set lru3[(N-1):0] to decimal 1 and lru2[(N-1):0] to decimal 0
        elseif(lru2[(N-1):0] corresponding to the entry is greater than lru0[(N-1):0] and lru3[(N-1):0])
            Set lru2[(N-1):0] to decimal 2
            if(lru0[(N-1):0] corresponding to the entry is greater than lru3[(N-1):0])
                Set lru0[(N-1):0] to decimal 1 and lru3[(N-1):0] to decimal 0
            else Set lru3[(N-1):0] to decimal 1 and lru0[(N-1):0] to decimal 0
```

Table 9: Full LRU Algorithm as Applied to a 4-Way Set Associative Cache Memory

```
        else Set lru3[(N-1):0] corresponding to the entry to decimal 2
          if(lru0[(N-1):0] corresponding to the entry is greater than lru2[(N-1):0])
             Set lru0[(N-1):0] to decimal 1 and lru2[(N-1):0] to decimal 0
          else Set lru2[(N-1):0] to decimal 1 and lru0[(N-1):0] to decimal 0
      else
      Increment the N-bit Counter corresponding to the entry by 1 while you are updating the entry
      Set lru1[(N-1):0] for the entry in Way1 with the incremented value of N-bit Counter
      lru vectors corresponding to Way0, Way2 and Way3 i.e lru0[(N-1):0], lru2[(N-1):0] and lru3[(N-1):0] remain unchanged
  elseif(lru2[(N-1):0] has the least count compared to other lru vectors corresponding to the indexed entry)
  Replace entry in Way2
  if(N-bit counter corresponding to the entry which had a MISS is at its max count)
     Set N-bit counter and lru2[(N-1):0] to decimal 3 while you are updating the entry in Way2
     if(lru0[(N-1):0] corresponding to the entry is greater than lru1[(N-1):0] and lru3[(N-1):0])
        Set lru0[(N-1):0] to decimal 2
        if(lru1[(N-1):0] corresponding to the entry is greater than lru3[(N-1):0])
           Set lru1[(N-1):0] to decimal 1 and lru3[(N-1):0] to decimal 0
        else Set lru3[(N-1):0] to decimal 1 and lru1[(N-1):0] to decimal 0
     elseif(lru1[(N-1):0] corresponding to the entry is greater than lru0[(N-1):0] and lru3[(N-1):0])
        Set lru1[(N-1):0] to decimal 2
        if(lru0[(N-1):0] corresponding to the entry is greater than lru3[(N-1):0])
           Set lru0[(N-1):0] to decimal 1 and lru3[(N-1):0] to decimal 0
        else Set lru3[(N-1):0] to decimal 1 and lru0[(N-1):0] to decimal 0
     else Set lru3[(N-1):0] corresponding to the entry to decimal 2
        if(lru0[(N-1):0] corresponding to the entry is greater than lru1[(N-1):0])
           Set lru0[(N-1):0] to decimal 1 and lru1[(N-1):0] to decimal 0
        else Set lru1[(N-1):0] to decimal 1 and lru0[(N-1):0] to decimal 0
  else
  Increment the N-bit Counter corresponding to the entry by 1 while you are updating the entry
  Set lru2[(N-1):0] for the entry in Way2 with the incremented value of N-bit Counter
  lru vectors corresponding to Way0, Way1 and Way3 i.e lru0[(N-1):0], lru1[(N-1):0] and lru3[(N-1):0] remain unchanged

else
  Replace entry in Way3
  if(N-bit counter corresponding to the entry which had a MISS is at its max count)
     Set N-bit counter and lru3[(N-1):0] to decimal 3 while you are updating the entry in Way3
     if(lru0[(N-1):0] corresponding to the entry is greater than lru1[(N-1):0] and lru2[(N-1):0])
        Set lru0[(N-1):0] to decimal 2
        if(lru1[(N-1):0] corresponding to the entry is greater than lru2[(N-1):0])
           Set lru1[(N-1):0] to decimal 1 and lru2[(N-1):0] to decimal 0
        else Set lru2[(N-1):0] to decimal 1 and lru1[(N-1):0] to decimal 0
     elseif(lru1[(N-1):0] corresponding to the entry is greater than lru0[(N-1):0] and lru2[(N-1):0])
        Set lru1[(N-1):0] to decimal 2
        if(lru0[(N-1):0] corresponding to the entry is greater than lru2[(N-1):0])
           Set lru0[(N-1):0] to decimal 1 and lru2[(N-1):0] to decimal 0
        else Set lru2[(N-1):0] to decimal 1 and lru0[(N-1):0] to decimal 0
     else Set lru2[(N-1):0] corresponding to the entry to decimal 2
        if(lru0[(N-1):0] corresponding to the entry is greater than lru1[(N-1):0])
           Set lru0[(N-1):0] to decimal 1 and lru1[(N-1):0] to decimal 0
        else Set lru1[(N-1):0] to decimal 1 and lru0[(N-1):0] to decimal 0
  else
  Increment the N-bit Counter corresponding to the entry by 1 while you are updating the entry
  Set lru3[(N-1):0] for the entry in Way3 with the incremented value of N-bit Counter
  lru vectors corresponding to Way0, Way1 and Way2 i.e lru0[(N-1):0], lru1[(N-1):0] and lru2[(N-1):0] remain unchanged
```

15. Describe the way Round Robin Algorithm gets used in replacing an entry in a 4-Way Set Associative Cache Memory?

Table below shows the way Round Robin Algorithm gets used in replacing an entry in a 4-Way Set Associative Cache Memory.

Table 10: Round Robin Algorithm as Applied to a 4-Way Set Associative Cache Memory

1. Maintain a 2-bit Round Robin array with number of entries equal to the number of entries in the Tag array as shown in the figure below. In figure below the bits used to index the Data and Tag array may be same or different (i.e index0 = index 1 or index0 != index1).

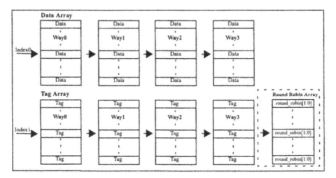

Figure 32: RR for a 4-Way Set Associative Cache Memory

2. At Power on, reset the entire Round Robin Array to all 0's.

3. Round Robin array remains untouched in the case of a Cache Hit.

4. Round Robin array remains untouched in the case of a Snoop Invalidate.

5. Use the following algorithm to replace an entry and update the Round Robin array in the case of a Cache Miss.

```
if(round_robin[1:0] for the indexed entry is 2'b00)
    Replace entry in Way0
    Set round_robin[1:0] to 2'b01 while you are updating the entry in Way0
elseif(round_robin[1:0] for the indexed entry is 2'b01)
    Replace entry in Way1
    Set round_robin[1:0] to 2'b10 while you are updating the entry in Way1
elseif(round_robin[1:0] for the indexed entry is 2'b10)
    Replace entry in Way2
    Set round_robin[1:0] to 2'b11 while you are updating the entry in Way2
else(round_robin[1:0] for the indexed entry is 2'b11)
    Replace entry in Way3
    Set round_robin[1:0] to 2'b00 while you are updating the entry in Way3
```

16. What do you mean by Coherency and what are the various Cache Coherency Protocols used?

Coherency problem refers to inconsistency of distributed cached copies of the same cache line addressed from the shared memory. A Memory System is Coherent if it meets the following three requirements -

1. Write to a location by processor 'P' followed by a read to the same location by processor 'P' returns the value written by processor 'P' as long as there are no other writes by other processors to that location in between the write and read of processor 'P' i.e In figure below if CPU 1 writes 'A' to location 'Y' then all future reads of location 'Y' will return 'A' if no other processor writes to location 'Y' after CPU 1.

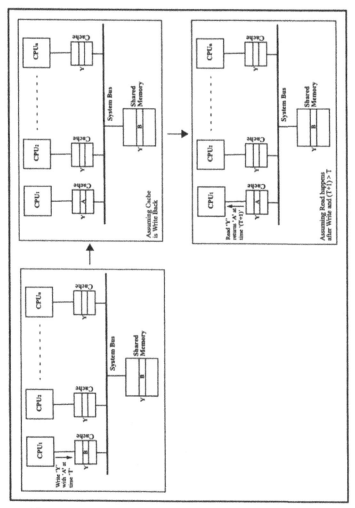

Figure 33: Requirement 1 for a Coherent Memory System

2. Write to a location by processor 'P' eventually gets seen by other processors making a read to the same location as long as the write and read are sufficiently separated and there are no other writes happening to that location by other processors in

between the write and read i.e In figure below if CPU 1 writes 'A' to location 'Y', CPU 2 will eventually be able to read value 'A' from location 'Y' as long as there are no other writes to location 'Y' in between the write made by CPU 1 and the read made by CPU 2.

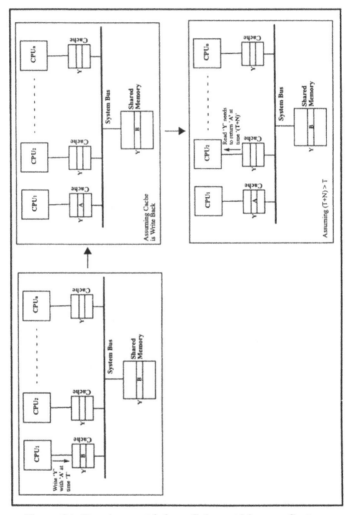

Figure 34: Requirement 2 for a Coherent Memory System

3. Writes to the same location are serialized i.e In figure below if CPU's 1 and 2 both write to location 'Y', all processors see the same order of writes.

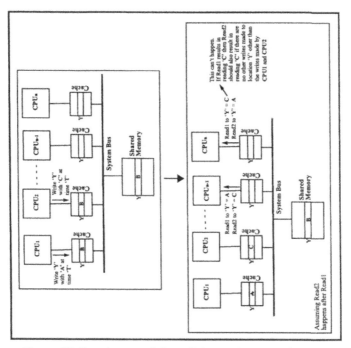

Figure 35: Requirement 3 for a Coherent Memory System

Protocol is a means by which caches, processors, main memory and bus masters communicate with each other. Cache Coherency protocol is a means by which all caches within a system assure that coherency is maintained and bus collisions do not occur.

Table below shows the various schemes used to maintain Coherency.

Table 11: Schemes used to Maintain Coherency

Scheme	Description
Software Scheme	This scheme generally depends on the actions of the programmer, compiler or the operating system in dealing with the coherence problem. Some of the methods used here are 1. declaring shared data as non-cacheable. 2. allowing caching of shared data and providing some special cache managing instructions for cache flush or selective invalidation in order to maintain coherence. Decision about coherence related actions are often made statically (i.e while coding if programmer and during compiler analysis if compiler) if we were relying on the actions of the programmer or compiler to maintain coherency.

Table 11: Schemes used to Maintain Coherency

Scheme	Description
	This scheme yields better performance if the amount of shared data is limited but if the processors are cooperatively working on a common application sharing a large database then degradations due to memory access are likely to be felt. Software schemes are generally less expensive than their hardware counterparts but their inefficiencies in maintaining coherency makes their usage less favorable when compared to their hardware counterparts.
Hardware Scheme	This scheme deals with the coherence problem by dynamic recognition of inconsistency conditions for shared data entirely at run time. This scheme promises better performance when compared to software scheme. Being totally transparent to software, hardware scheme frees the programmer and compiler from any responsibility for coherence maintenance. Two basic hardware schemes used in the industry are *Directory (Point to Point) based Scheme* In this scheme the global, system wide status information relevant for coherence maintenance is stored in some kind of a central or distributed directory. Here directories can be organized in different ways and it is the responsibility of the central or distributed memory controller to take appropriate actions to preserve the coherence by sending directed (point-to-point) individual messages to known locations, avoiding the broadcasts. This scheme is primarily suitable for large scale multiprocessor systems with interconnection networks. *Broadcast (Snoop) Scheme* Unlike the directory based scheme, in this scheme the responsibility for maintaining coherence is distributed among local caches. The name (i.e snoop) comes from the ability of the local cache controllers to snoop on the shared bus, while all processors have to broadcast their requests that can modify the coherence state of shared blocks. The importance of this scheme comes from its relative simplicity, low-cost implementation and ease in system expansion. Since all processors share a common bus, bus traffic puts an upper limit on the number of processors that can be supported. Because of the bus traffic this scheme is typically used in systems with a small or medium number of processors. Two write policies usually applied in this scheme are write-invalidate and write-update/write-broadcast.

Tables 13 and 14 below summarizes MOSI (as applied to Inclusive, Non-CMP Multi-Processor System shown in Figure 37), MOESI (as applied to Inclusive, Non-CMP MultiProcessor System shown in Figure 37), MHOSI (as applied to Inclusive, CMP MultiProcessor System shown in Figure 38) and MEI (as applied to Inclusive, Non-CMP UniProcessor System shown in Figure 36) Cache Coherency Protocols as used in the industry.

System configurations (i.e Non-CMP, UniProcessor System, Non-CMP MultiProcessor System and CMP MultiProcessor System) and State definitions used in Tables 13 and 14 are described in Table 12 below.

Table 12: System Configuration Description and State Definitions

System Configurations	
Non-CMP, UniProcessor System	Figure below shows a Non-CMP UniProcessor system. 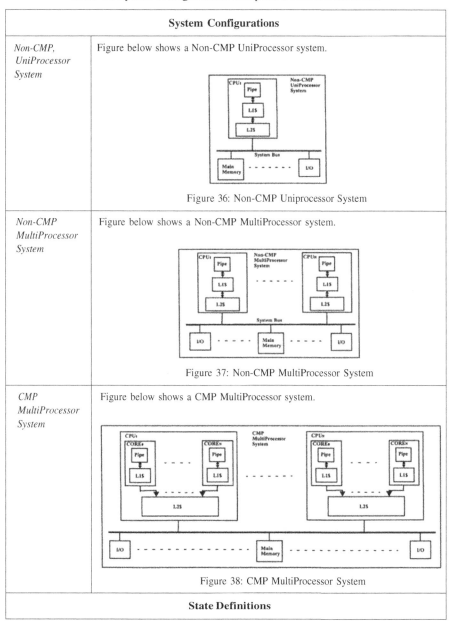 Figure 36: Non-CMP Uniprocessor System
Non-CMP MultiProcessor System	Figure below shows a Non-CMP MultiProcessor system. Figure 37: Non-CMP MultiProcessor System
CMP MultiProcessor System	Figure below shows a CMP MultiProcessor system. Figure 38: CMP MultiProcessor System
State Definitions	

Table 12: System Configuration Description and State Definitions

Definition of various states in MEI, MOSI and MOESI	**M** - Line is modified by this CPU and is not shared by other CPU's within the System and the line may or may not be present in lower level Caches within this CPU. **O** - Line is modified by this CPU and is shared by other CPU's within the System and the line may or may not be present in lower level Caches within this CPU. **E** - Line is unmodified by this CPU and is not shared by other CPU's within the System and the line may or may not be present in lower level Caches within this CPU. **S** - Line is unmodified by this CPU and is shared by other CPU's within the System and the line may or may not be present in lower level Caches within this CPU. **I** - Line is Invalid within this CPU.
Definition of various states in MHOSI	**M** - Line is modified by this CPU and is not shared by other CPU's within the System and the line is not present in any of the lower level Caches within this CPU. **H** - Line is modified by this CPU and is not shared by other CPU's within the System and the line is present in at least one of the lower level Caches within this CPU. **O** - Line is modified by this CPU and is shared by other CPU's within the System and the line may or may not be present in lower level Caches within this CPU. **S** - Line is unmodified by this CPU and is shared by other CPU's within the System and the line may or may not be present in lower level Caches within this CPU. **I** - Line is Invalid within this CPU.

For the System Configurations shown in Figures 36, 37 and 38, following things are assumed

1. L1$ is write through whereas L2$ is write back.
2. Stores do not allocate on a store miss in L1$.
3. Coherency is maintained in L2$.
4. L2$ is inclusive (i.e at any given time data sitting in L1$ is a subset of L2$).

In Tables 13 and 14 below, Hit is a condition where you have an address match and the Line is in a state other than I (Invalid) state.

Table 13: MOSI, MOESI and MHOSI Cache Coherency Protocols

Operation	MOSI (Inclusive, Non-CMP Multi-Processor)		MOESI (Inclusive, Non-CMP Multi-Processor)		MHOSI (Inclusive, CMP Multi-Processor)	
	CS	Operation, NS	CS	Operation, NS	CS	Operation, NS
Local Read Hit	M	1. Read data. 2. Next State = **M**	M	1. Read data. 2. Next State = **M**	M	1. Read data. 2. Next State = **H**
	O	1. Read Data. 2. Next State = **O**	O	1. Read Data. 2. Next State = O	H	1. Read Data. 2. Next State = **H**
	S	1. Read Data. 2. Next State = **S**	E	1. Read Data. 2. Next State = **E**	O	1. Read Data. 2. Next State = O
	I	-	S	1. Read Data. 2. Next State = **S**	S	1. Read Data. 2. Next State = **S**
			I	-	I	-
Local Read Miss	M	1. Evict data to Main memory and Invalidate local Caches which has data corresponding to the evicted address. 2. Place Read-to-Share request over the System bus which forces Main Memory or foreign Caches over the System bus to forward the latest data corresponding to the Read Miss address. 3. Update entry with required data from Main memory or foreign Caches. 4. Next State = **S**	M	1. Evict data to Main memory and Invalidate local Caches which has data corresponding to the evicted address. 2. Place Read-to-Share request over the System bus which forces Main Memory or foreign Caches over the System bus to forward the latest data corresponding to the Read Miss address. 3. Update entry with required data from Main memory or foreign Caches over the System bus. 4. Next State = **E** if Main memory forwarded the data and it is the only one which has it **or** Next State = **S** if the data was forwarded by foreign Cache over the System bus.	M	1. Evict data to Main memory. 2. Place Read-to-Share request over the System bus which forces Main Memory or foreign Caches over the System bus to forward the latest data corresponding to the Read Miss address. 3. Update entry with required data from Main memory or foreign Caches. 4. Next State = **S**

Table 13: MOSI, MOESI and MHOSI Cache Coherency Protocols

Operation	MOSI (Inclusive. Non-CMP Multi-Processor)		MOESI (Inclusive, Non-CMP Multi-Processor)		MHOSI (Inclusive, CMP Multi-Processor)	
	CS	Operation, NS	CS	Operation, NS	CS	Operation, NS
	O	1. Evict data to Main memory and Invalidate local Caches which has data corresponding to the evicted address. 2. Place Read-to-Share request over the System bus which forces Main Memory or foreign Caches over the System bus to forward the latest data corresponding to the Read Miss address. 3. Update entry with required data from Main memory or foreign Caches over the System bus. 4. Next State = S	O	1. Evict data to Main memory and Invalidate local Caches which has data corresponding to the evicted address. 2. Place Read-to-Share request over the System bus which forces Main Memory or foreign Caches over the System bus to forward the latest data corresponding to the Read Miss address. 3. Update entry with required data from Main memory or foreign Caches over the System bus. 4. Next State = E if Main memory forwarded the data and it is the only one which has it or Next State = S if the data was forwarded by foreign Cache over the System bus.	H	1. Evict data to Main memory and Invalidate local Caches which has data corresponding to the evicted address. 2. Place Read-to-Share request over the System bus which forces Main Memory or foreign Caches over the System bus to forward the latest data corresponding to the Read Miss address. 3. Update entry with required data from Main memory or foreign Caches over the System bus. 4. Next State = S
	S	1. Invalidate local Caches which has data corresponding to the Replaced address. 2. Place Read-to-Share request over the System bus which forces Main Memory or foreign Caches over the System bus to forward the latest data corresponding to the Read Miss address. 3. Update entry with required data from Main memory or foreign Caches over the System bus. 4. Next State = S	E	1. Invalidate local Caches which has data corresponding to the Replaced address. 2. Place Read-to-Share request over the System bus which forces Main Memory or foreign Caches over the System bus to forward the latest data corresponding to the Read Miss address. 3. Update entry with required data from Main memory or foreign Caches over the System bus. 4. Next State = E if Main memory forwarded the data and it is the only one which has it or Next State = S if the data was forwarded by foreign Cache over the System bus.	O	1. Evict data to Main memory and Invalidate local Caches which has data corresponding to the evicted address. 2. Place Read-to-Share request over the System bus which forces Main Memory or foreign Caches over the System bus to forward the latest data corresponding to the Read Miss address. 3. Update entry with required data from Main memory or foreign Caches over the System bus. 4. Next State = S

Table 13: MOSI, MOESI and MHOSI Cache Coherency Protocols

Operation	MOSI (Inclusive, Non-CMP Multi-Processor)		MOESI (Inclusive, Non-CMP Multi-Processor)		MHOSI (Inclusive, CMP Multi-Processor)	
	CS	Operation, NS	CS	Operation, NS	CS	Operation, NS
	I	1. Place Read-to-Share request over the System bus which forces Main Memory or foreign Caches over the System bus to forward the latest data corresponding to the Read Miss address. 2. Update entry with required data from Main memory or foreign Caches over the System bus. 3. Next State = S	S	1. Invalidate local Caches which has data corresponding to the Replaced address. 2. Place Read-to-Share request over the System bus which forces Main Memory or foreign Caches over the System bus to forward the latest data corresponding to the Read Miss address. 3. Update entry with required data from Main memory or foreign Caches over the System bus. 4. Next State = E if Main memory forwarded the data and it is the only one which has it or Next State = S if the data was forwarded by foreign Cache over the System bus.	S	1. Invalidate local Caches which has data corresponding to the Replaced address. 2. Place Read-to-Share request over the System bus which forces Main Memory or foreign Caches over the System bus to forward the latest data corresponding to the Read Miss address, 3. Update entry with required data from Main memory or foreign Caches over the System bus. 4. Next State = S
			I	1. Place Read-to-Share request over the System bus which forces Main Memory or foreign Caches over the System bus to forward the latest data corresponding to the Read Miss address. 2. Update entry with required data from Main memory or foreign Caches over the System bus. 3.Next State = E if Main memory forwarded the data and it is the only one which has it or Next State = S if the data was forwarded by foreign Cache over the System bus.	I	1. Place Read-to-Share request over the System bus which forces Main Memory or foreign Caches over the System bus to forward the latest data corresponding to the Read Miss address. 2. Update entry with required data from Main memory or foreign Caches over the System bus. 3. Next State = S

Table 13: MOSI, MOESI and MHOSI Cache Coherency Protocols

Operation	MOSI (Inclusive, Non-CMP Multi-Processor)		MOESI (Inclusive, Non-CMP Multi-Processor)		MHOSI (Inclusive, CMP Multi-Processor)	
	CS	Operation, NS	CS	Operation, NS	CS	Operation, NS
Local Write Hit	M	1. Update entry with new data 2. Next State = M	M	1. Update entry with new data 2. Next State = M	M	1. Update entry with new data 2. Next State = M
	O	1. Place Write-to-Invalidate request over the System bus which forces foreign Caches to invalidate data corresponding to the Write Hit address. 2. Update entry with new data. 3. Next State = M	O	1. Place Write-to-Invalidate request over the System bus which forces foreign Caches to invalidate data corresponding to the Write Hit address. 2. Update entry with new data. 3. Next State = M	H	1. Invalidate local Caches (excluding the one which placed the Write request) which has data corresponding to the Write Hit address. 2. Update entry with new data. 3. Next State = M if the cache which placed the request had a Miss for the Write address **or** Next State = H if the cache which placed the request had a Hit for the Write address.
	S	1. Place Write-to-Invalidate request over the System bus which forces foreign Caches to invalidate data corresponding to the Write Hit address. 2. Update entry with new data. 3. Next State = M	E	1. Update entry with new data 2. Next State = M	O	1. Invalidate local Caches (excluding the one which placed the Write request) which has data corresponding to the Write Hit address. 2. Place Write-to-Invalidate request over the System bus which forces foreign Caches to invalidate data corresponding to the Write Hit address. 3. Update entry with new data. 4. Next State = M if the cache which placed the request had a Miss for the Write address **or** Next State = H if the cache which placed the request had a Hit for the Write address.

Table 13: MOSI, MOESI and MHOSI Cache Coherency Protocols

Operation	MOSI (Inclusive, Non-CMP Multi-processor)		MOESI (Inclusive, Non-CMP Multi-processor)		MHOSI (Inclusive, CMP Multi-Processor)	
	CS	Operation, NS	CS	Operation, NS	CS	Operation, NS
	I	-	S	1. Place Write-to-Invalidate request over the System bus which forces foreign Caches to invalidate data corresponding to the Write Hit address. 2. Update entry with new data. 3. Next State = M	S	1. Invalidate local Caches (excluding the one which placed the Write request) which has data corresponding to the Write Hit address. 2. Place Write-to-Invalidate request over the System bus which forces foreign Caches to invalidate data corresponding to the Write Hit address. 3. Update entry with new data. 4. Next State = M if the cache which placed the request had a Miss for the Write address **or** Next State =H if the cache which placed the request had a Hit for the Write address.
			I	-	I	-
Local Write Miss	M	1. Evict data to Main memory and Invalidate local Caches which has data corresponding to the Evicted address. 2. Place Read-to-Own request over the System bus which forces Main Memory or foreign Caches over the System bus to forward the latest data corresponding to the Write Miss address and also forces foreign Caches to invalidate data corresponding to the Write Miss address. 3. Update entry with data formed by merging data (i.e either from foreign Caches or Main Memory) over System bus with the latest data. 4. Next State = M	M	1. Evict data to Main memory and Invalidate local Caches which has data corresponding to the Evicted address. 2. Place Read-to-Own request over the System bus which forces Main Memory or foreign Caches over the System bus to forward the latest data corresponding to the Write Miss address and also forces foreign Caches to invalidate data corresponding to the Write Miss address. 3. Update entry with data formed by merging data (i.e either from foreign Caches or Main Memory) over System bus with the latest data. 4. Next State = M	M	1. Evict data to Main memory. 2. Place Read-to-Own request over the System bus which forces Main Memory or foreign Caches over the System bus to forward the latest data corresponding to the Write Miss address and also forces foreign Caches to invalidate data corresponding to the Write Miss address. 3. Update entry with data formed by merging data (i.e either from foreign Caches or Main Memory) over System bus with the latest data. 4. Next State = M

Table 13: MOSI, MOESI and MHOSI Cache Coherency Protocols

Operation	MOSI (Inclusive, Non-CMP Multi-Processor)		MOESI (Inclusive, Non-CMP Multi-Processor)		MHOSI (Inclusive, CMP Multi-Processor)	
	CS	Operation, NS	CS	Operation, NS	CS	Operation, NS
	O	1. Evict data to Main memory and Invalidate local Caches which has data corresponding to the Evicted address. 2. Place Read-to-Own request over the System bus which forces Main Memory or foreign Caches over the System bus to forward the latest data corresponding to the Write Miss address and also forces foreign Caches to invalidate data corresponding to the Write Miss address. 3. Update entry with data formed by merging data (i.e either from foreign Caches or Main Memory) over System bus with the latest data. 4. Next State = **M**	O	1. Evict data to Main memory and Invalidate local Caches which has data corresponding to the Evicted address. 2. Place Read-to-Own request over the System bus which forces Main Memory or foreign Caches over the System bus to forward the latest data corresponding to the Write Miss address and also forces foreign Caches to invalidate data corresponding to the Write Miss address. 3. Update entry with data formed by merging data (i.e either from foreign Caches or Main Memory) over System bus with the latest data. 4. Next State = **M**	H	1. Evict data to Main memory and Invalidate local Caches which has data corresponding to the Evicted address. 2. Place Read-to-Own request over the System bus which forces Main Memory or foreign Caches over the System bus to forward the latest data corresponding to the Write Miss address and also forces foreign Caches to invalidate data corresponding to the Write Miss address. 3. Update entry with data formed by merging data (i.e either from foreign Caches or Main Memory) over System bus with the latest data. 4. Next State = **M**
	S	1. Invalidate local Caches which has data corresponding to the Replaced address. 2. Place Read-to-Own request over the System bus which forces Main Memory or foreign Caches over the System bus to forward the latest data corresponding to the Write Miss address and also forces foreign Caches to invalidate data corresponding to the Write Miss address. 3. Update entry with data formed by merging data (i.e either from foreign Caches or Main Memory) over System bus with the latest data. 4. Next State = **M**	E	1. Invalidate local Caches which has data corresponding to the Replaced address. 2. Place Read-to-Own request over the System bus which forces Main Memory or foreign Caches over the System bus to forward the latest data corresponding to the Write Miss address and also forces foreign Caches to invalidate data corresponding to the Write Miss address. 3. Update entry with data formed by merging data (i.e either from foreign Caches or Main Memory) over System bus with the latest data. 4. Next State = **M**	O	1. Evict data to Main memory and Invalidate local Caches which has data corresponding to the Evicted address. 2. Place Read-to-Own request over the System bus which forces Main Memory or foreign Caches over the System bus to forward the latest data corresponding to the Write Miss address and also forces foreign Caches to invalidate data corresponding to the Write Miss address. 3. Update entry with data formed by merging data (i.e either from foreign Caches or Main Memory) over System bus with the latest data. 4. Next State = **M**

Table 13: MOSI, MOESI and MHOSI Cache Coherency Protocols

Operation	MOSI (Inclusive, Non-CMP Multi-Processor)		MOESI (Inclusive, Non-CMP Multi-Processor)		MHOSI (Inclusive, CMP Multi-Processor)	
	CS	Operation, NS	CS	Operation, NS	CS	Operation, NS
	I	1. Place Read-to-Own request over the System bus which forces Main Memory or foreign Caches over the System bus to forward the latest data corresponding to the Write Miss address and also forces foreign Caches to invalidate data corresponding to the Write Miss address. 2. Update entry with data formed by merging data (i.e either from foreign Caches or Main Memory) over System bus with the latest data. 3. Next State = M	S	1. Invalidate local Caches which has data corresponding to the Replaced address. 2. Place Read-to-Own request over the System bus which forces Main Memory or foreign Caches over the System bus to forward the latest data corresponding to the Write Miss address and also forces foreign Caches to invalidate data corresponding to the Write Miss address. 3. Update entry with data formed by merging data (i.e either from foreign Caches or Main Memory) over System bus with the latest data. 4. Next State = M	S	1. Invalidate local Caches which has data corresponding to the Replaced address. 2. Place Read-to-Own request over the System bus which forces Main Memory or foreign Caches over the System bus to forward the latest data corresponding to the Write Miss address and also forces foreign Caches to invalidate data corresponding to the Write Miss address. 3. Update entry with data formed by merging data (i.e either from foreign Caches or Main Memory) over System bus with the latest data. 4. Next State = M
			I	1. Place Read-to-Own request over the System bus which forces Main Memory or foreign Caches over the System bus to forward the latest data corresponding to the Write Miss address and also forces foreign Caches to invalidate data corresponding to the Write Miss address. 2. Update entry with data formed by merging data (i.e either from foreign Caches or Main Memory) over System bus with the latest data. 3. Next State = M	I	1. Place Read-to-Own request over the System bus which forces Main Memory or foreign Caches over the System bus to forward the latest data corresponding to the Write Miss address and also forces foreign Caches to invalidate data corresponding to the Write Miss address. 2. Update entry with data formed by merging data (i.e either from foreign Caches or Main Memory) over System bus with the latest data. 3. Next State = M

Table 13: MOSI, MOESI and MHOSI Cache Coherency Protocols

Operation	MOSI (Inclusive, Non-CMP Multi-processor)		MOESI (Inclusive, Non-CMP Multi-processor)		MHOSI (Inclusive, CMP Multi-Processor)	
	CS	Operation, NS	CS	Operation, NS	CS	Operation, NS
Snoop Hit (Read-to-Share Request)	M	1. Provide data. 2. Next State = O	M	1. Provide data. 2. Next State = O	M	1. Provide data. 2. Next State = O
	O	1. Provide Data. 2. Next State = O	O	1. Provide Data. 2. Next State = O	H	1. Provide Data. 2. Next State = O
	S	1. Next State = S	E	1. Provide data. 2. Next State = S	O	1. Provide Data. 2. Next State = O
	I	-	S	1. Next State = S	S	1. Next State = S
			I	-	I	-
Snoop Hit (Read-to-Own Request)	M	1. Provide data. 2. Invalidate local Caches which has data corresponding to the Snoop address. 3. Next State = I	M	1. Provide data. 2. Invalidate local Caches which has data corresponding to the Snoop address. 3. Next State = I	M	1. Provide data. 2. Next State =I
	O	1. Provide data. 2. Invalidate local Caches which has data corresponding to the Snoop address. 3. Next State = I	O	1. Provide data. 2. Invalidate local Caches which has data corresponding to the Snoop address. 3. Next State = I	H	1. Provide data. 2. Invalidate local Caches which has data corresponding to the Snoop address. 3. Next State = I
	S	1. Invalidate local Caches which has data corresponding to the Snoop address. 2. Next State = I	E	1. Provide data. 2. Invalidate local Caches which has data corresponding to the Snoop address. 3. Next State =I	O	1. Provide data. 2. Invalidate local Caches which has data corresponding to the Snoop address. 3. Next State = I
	I	-	S	1. Invalidate local Caches which has data corresponding to the Snoop address. 2. Next State = I	S	1. Invalidate local Caches which has data corresponding to the Snoop address. 2. Next State = I
			I	-	I	-

Table 13: MOSI, MOESI and MHOSI Cache Coherency Protocols

Operation	MOSI (Inclusive, Non-CMP Multi-Processor)		MOESI (Inclusive, Non-CMP Multi-processor)		MHOSI (Inclusive, CMP Multi-Processor)	
	CS	Operation, NS	CS	Operation, NS	CS	Operation, NS
Snoop Hit (Write-to Invalidate Request)	M	-	M	-	M	-
	O	1. Invalidate local Caches which has data corresponding to the Snoop address. 2. Next State = **I**	O	1. Invalidate local Caches which has data corresponding to the Snoop address. 2. Next State = **I**	H	-
	S	1. Invalidate local Caches which has data corresponding to the Snoop address. 2. Next State = **I**	E	-	O	1. Invalidate local Caches which has data corresponding to the Snoop address. 2. Next State = **I**
	I	-	S	1. Invalidate local Caches which has data corresponding to the Snoop address. 2. Next State = I	S	1. Invalidate local Caches which has data corresponding to the Snoop address. 2. Next State = **I**
			I	-	I	-

Table below summarizes MEI Cache Coherency Protocol as applied to a inclusive, Non-CMP Uniprocessor System.

Table 14: MEI Cache Coherency Protocol

	MEI (Inclusive, Non-CMP Uniprocessor)			
CS	Local Read Hit (Operation, NS)	Local Read Miss (Operation, NS)	Local Write Hit (Operation, NS)	Local Write Miss (Operation, NS)
M	1. Read data. 2. Next State = **M**	1. Evict data to Main memory and Invalidate local Caches which has data corresponding to the evicted address. 2. Place Read request to Main memory. 3. Update entry with required data from Main memory, 4. Next State = **E**	1. Update entry with new data 2. Next State = **M**	1. Evict data to Main memory and Invalidate local Caches which has data corresponding to the Evicted address. 2. Place Read request to Main memory. 3. Update entry with data formed by merging data from Main Memory with the latest data. 4. Next State = **M**

Table 14: MEI Cache Coherency Protocol

MEI			
(Inclusive, Non-CMP Uniprocessor)			
CS	**Local Read Hit** (Operation, NS)	**Local Read Miss** (Operation, NS)	**Local Write Hit** (Operation, NS)	**Local Write Miss** (Operation, NS)
E	1. Read Data. 2. Next State = **E**	1. Invalidate local Caches which has data corresponding to the Replaced address. 2. Place Read request to Main memory. 3. Update entry with required data from Main memory. 4. Next State = **E**	1. Update entry with new data 2. Next State = **M**	1. Invalidate local Caches which has data corresponding to the Replaced address. 2. Place Read request to Main memory. 3. Update entry with data formed by merging data from Main Memory with the latest data. 4. Next State = **M**
I	-	1. Place Read request to Main memory. 2. Update entry with required data from Main memory. 3. Next State = **E**	-	1. Place Read request to Main memory. 2. Update entry with data formed by merging data from Main Memory with the latest data. 3. Next State = **M**

	I/O Snoop Read Hit (Operation, NS)	**I/O Snoop Read Miss** (Operation, NS)	**I/O Snoop Write Hit** (Operation, NS)	**I/O Snoop Write Miss** (Operation, NS)
M	1. Provide data. 2. Next State = **M**	-	1. Invalidate local Caches which has data corresponding to the Snoop address. 2. Update entry with new data 3. Next State = **M**	-
E	1. Provide data. 2. Next State = **E**	-	1. Invalidate local Caches which has data corresponding to the Snoop address. 2. Update entry with new data 3. Next State = **M**	-
I	-	-	-	-

17. Explain Non-Pipelining, Pipelining, Superscalar, In-Order Execution and Out-Of-Order Execution as applied to a Processor?

Table 15 below gives a brief description for Non-Pipelining, Pipelining, Superscalar, In-Order Execution and Out-Of-Order Execution as applied to a Processor.

Table 15: Non-Pipelining, Pipelining, Superscalar, In-Order and Out-Of-Order

Non-Pipelining	Here the Processor waits for an instruction to complete before it feeds a new instruction in the pipe (i.e there is no overlap in the instruction flow). Figure below illustrates the instruction flow in a Non-Pipelined Processor. *(Here we have assumed a 5-stage pipeline (i.e F (Fetch), D (Decode), R (Read Register File), E (Execute), W (Writeback)) where the pipes are separated by flops or latches.)*
Pipelining	Here the Processor can potentially feed a new instruction in the pipe every cycle by overlapping instruction execution. Figure below illustrates the instruction flow in a Pipelined Processor. *(Here we have assumed a 5-stage pipeline (i.e F (Fetch), D (Decode), R (Read Register File), E (Execute), W (Writeback)) where the pipes are separated by flops or latches.)*
Superscalar	Here the Processor executes more than one instruction in a pipe stage. Figure below illustrates the instruction flow in a Pipelined, 2-Way Superscalar Processor. *(Here we have assumed a 5-stage pipeline (i.e F (Fetch), D (Decode), R (Read Register File), E (Execute), W (Writeback)) where the pipes are separated by flops or latches.)*

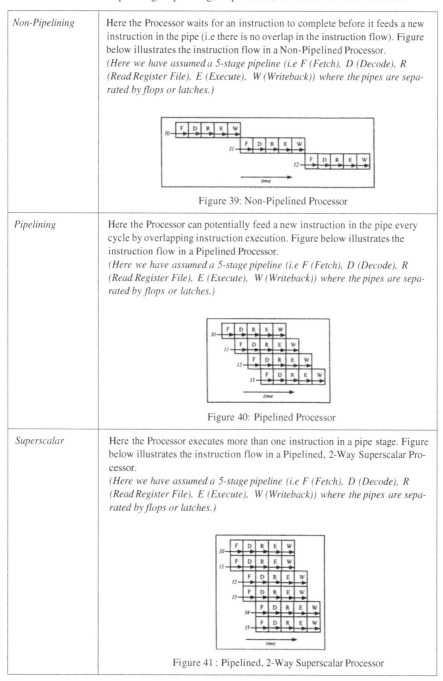

Figure 39: Non-Pipelined Processor

Figure 40: Pipelined Processor

Figure 41 : Pipelined, 2-Way Superscalar Processor

Table 15: Non-Pipelining, Pipelining, Superscalar, In-Order and Out-Of-Order

In-Order Execution	Here the Processor executes instructions in a sequential fashion i.e a younger instruction which is not dependent on any of the previous older instructions still has to wait for its execution until all the older instructions gets executed. Figure below illustrates In-Order Execution as applied to a Pipelined, 2-Way Superscalar Processor. 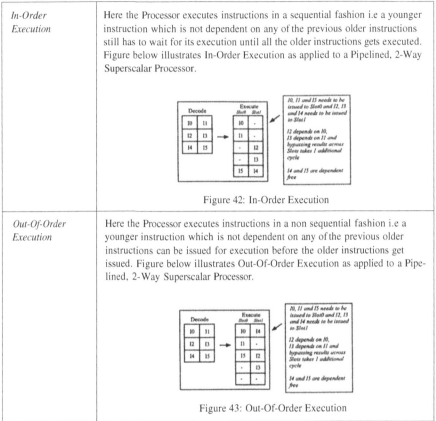 Figure 42: In-Order Execution
Out-Of-Order Execution	Here the Processor executes instructions in a non sequential fashion i.e a younger instruction which is not dependent on any of the previous older instructions can be issued for execution before the older instructions get issued. Figure below illustrates Out-Of-Order Execution as applied to a Pipelined, 2-Way Superscalar Processor. Figure 43: Out-Of-Order Execution

18. What is the equation for CPU Performance?

CPU performance is expressed as the amount of time it takes to execute a program. CPU performance is given by the following equation -

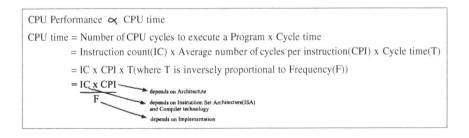

Table below provides Ideal CPU performance for various Implementations.

Table 16: Ideal CPU Performance

Implementation	Ideal CPU Performance (Perfect branch prediction with no Load/Store misses and Stalls) *(Assuming the following* *1. Program has 1000 instructions with 20% Branches, 20% Loads, 10% Stores and 50% ALU instructions* *2. Each instruction takes 1 Cycle to execute* *3. CPU operating frequency is 4GHz)*
Non-Pipelined *(assuming every instruction takes 5 cycles (i.e 5 pipe stages (F, D, R, E and W)) to execute)*	CPU time = IC x CPI/F $= (0.2CPI_B + 0.2CPI_L + 0.1CPI_S + 0.5CPI_{ALU}) \times 1000/4 \times 10^9$ $= (0.2x5 + 0.2x5 + 0.1x5 + 0.5x5) \times 250$ ns $= 5 \times 250$ ns $= 1250$ns
Pipelined, 1-Way Superscalar *(assuming we can retire only one instruction in a given cycle)*	CPU time = IC x CPI/F $= (0.2CPI_B + 0.2CPI_L + 0.1CPI_S + 0.5CPI_{ALU}) \times 1000/4 \times 10^9$ $= (0.2x1 + 0.2x1 + 0.1x1 + 0.5x1) \times 250$ ns $= 1 \times 250$ ns $= 250$ns
Pipelined, 3-Way Superscalar *(assuming we can retire any one of the following instruction groups in a given cycle -* *1. 3 ALU instructions or* *2. 1 Load instruction or* *3. 1 Store instruction or* *4. 3 Branch instructions)*	CPU time = IC x CPI/F $= (0.2CPI_B + 0.2CPI_L + 0.1CPI_S + 0.5CPI_{ALU}) \times 1000/4 \times 10^9$ $= (0.2x0.3 + 0.2x1 + 0.1x1 + 0.5x0.3) \times 250$ ns $= 0.51 \times 250$ ns $= 127.5$ns

19. Describe a simple Out-Of-Order CMP Chip with an example?

A typical CMP chip has more than one identical processor core, all of them connected to a Memory Subsystem (MS) as shown in Figure 44 below. A processor core in a CMP chip is implementation dependent i.e it could be an out-of-order or in-order core. The choice of implementation really depends on the kind of applications you are targeting for. An out-of-order processor core as shown in Figure 44 below typically has 6 major units which are Fetch Unit (FU), Decode Unit (DU), Rename and Issue Unit (RIU), Execution Unit (EXU), Dcache Unit (DCU) and Commit Unit (CU).

Typically FU is responsible for fetching instructions which need to be executed by the Core. It could have more than one pipe stage. In many cases it is also responsible for the following tasks - accessing ITLB (Instruction Translation Lookaside Buffer) for translating Virtual address to Physical address for instruction access, accessing I$ (Instruction Cache) for instructions, providing support for static/dynamic branch prediction, accessing DCU (assuming MMU (Memory Management Unit) is sitting in DCU) in the case of ITLB miss, accessing MS in the case of I$ miss, providing parity or ECC for the instructions and detecting few exception conditions (eg. parity error etc.).

Typically DU is responsible for decoding instructions. In many cases it is also responsible for renaming destination registers and managing various resources in various units down the pipeline. Typical resources being Commit Queue (queue which is responsible for handling in-order retirement of instructions), Issue Queue (queue where instructions go and sit waiting for them to be picked for issue), various Reorder Buffers or Working Register Files (temporary place holder for results before being written into Architectural Register File), Load Queue (queue responsible for handling Load instructions) and Store Queue (queue responsible for handling Store instructions).

Typically RIU is responsible for renaming source registers, picking instructions which are ready to be issued and issuing instructions by providing the necessary controls to the Architectural Register File (ARF), Working Register File (WRF) and Bypass Mux's.

Typically EXU is responsible for executing integer as well as floating point and graphics instructions. In many cases EXU has multiple similar functional units to support the superscalar nature of the pipe. In many cases it is also the place where architectural register files (ARF's) and working register files (WRF's) reside. In a 64-bit architecture these registers are 64-bit wide whereas in a 32-bit architecture they are 32-bits wide. In many cases in addition to data there could be some control information sitting in the WRF's.

Typically DCU is responsible for handling Loads and Stores. In many cases it has a Load Queue (LQ) for handling the Load instructions, Store Queue (SQ) for handling the Store instructions and a Data Cache (D$) for holding the most recently used data. It is also the place of residence for DTLB (Data TLB) and second level unified TLB (i.e MMU (Memory Management Unit)). DTLB is responsible for providing the address translation (i.e VA to PA) for Loads and Stores. MMU is responsible for providing address translations for the ones which missed in ITLB and DTLB. In some cases MMU could be represented as a separate unit in the block diagram.

Typically CU is responsible for committing instructions which have executed without exceptions. It commits instructions in-order by reading a value from WRF and writing it into ARF for the instructions which have valid destination registers and for the ones which doesn't have one it treats them as NOP's (i.e there is no movement of data into ARF). In many cases it is also responsible for handling external interrupts, internal exceptions/traps, redirecting FU in the case of branch misdirection and initializing PC and NPC coming out of RESET.

Typically MS is responsible for servicing requests that missed in Level 1 Caches of each of the individual cores and external Snoop requests. MS has a Crossbar Interface, additional levels of Caches (i.e L2, L3 etc.), on chip Memory Controllers and System Interface logic. MS could see one of the following requests from each core - D$ miss request, I$ miss request, MMU miss request (if supporting hardware table

walk), instruction prefetch from FU and data prefetch from DCU. MS is also responsible for maintaining cache coherency. Typical cache coherency protocols used are -
MEI, MOSI, MOESI and MHOSI

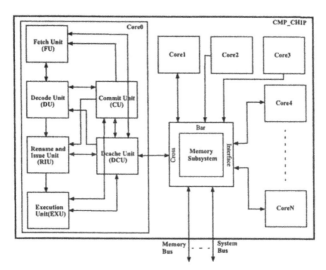

Figure 44: Block Diagram of a CMP Chip with an Out-Of-Order Processor Core

In the following paragraphs, tables and figures I tend to explain Core0 (an Out-Of-Order Processor Core) within the CMP Chip (shown in Figure 44 above) by taking an example.

Let's assume the Core to be a 64-bit, 3-Way superscalar Out-Of-Order Processor.
Let's assume a 10-stage Integer/Floating Point and Graphics pipeline and a 13-stage Memory pipeline which spans across various blocks within the Core as shown in Figure 45 below. Eventhough most of the floating point and graphics (fp&g) operations take more than one pipe stage, here for simplicity it's been assumed that these operations (including integer multiply and divide) take one pipe stage. The reason behind the number of pipe stages assumed for integer/fp&g/memory operations is to keep up with the current design trend of high frequency and high performance processors.

Table 17 below shows the functionality assumed within various blocks within the Core and Memory Subsystem. Here I have tried to be realistic by assuming the various cache sizes, queue depths and other functionality within the blocks to reflect the current design trend in the CPU design towards providing a good performance number for both commercial and technical applications. I have tried to provide enough functional information for each block to a point where it gets easy to understand the Out-Of-Orderliness of the Core. I need to warn you here that you may find some details missing for each block in the tables and figures below, this has been done on purpose to make it easier for explanation.

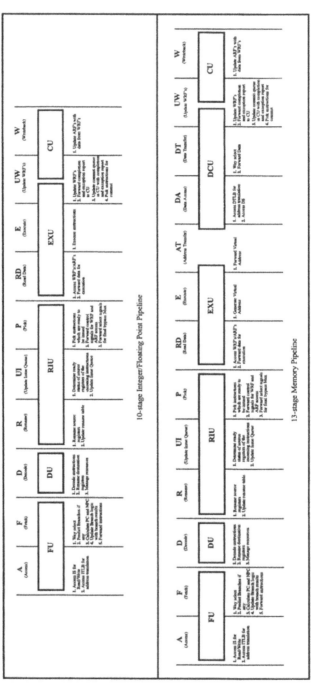

Figure 45: 10-Stage Integer/Floating Point & Graphics and 13-stage Memory Pipeline

Table 17: Functionality within Various Blocks

Block	Functionality
FU	It has a 32K, 8-Way, 1 Read port, 1 Write port I$, 64-entry fully associative translation lookaside buffer (FA-ITLB), a 2-bit Gshare branch prediction logic, a Branch target address buffer (BTA) and a 2-entry Instruction Miss Buffer (IMB). Assuming there is no self modifying code, there won't be any invalidation requests to I$. I$ here is responsible for storing instructions which has high probability of getting executed. FA-ITLB here is responsible for providing Physical Address for instruction access (for simplicity reasons I haven't shown context ID and other privileged and control bits as part of the FA-ITLB entry). 2-bit Gshare is responsible for providing the prediction information for branches. BTA is responsible for proving the target address for non-PC relative branches which are predicted as taken and IMB is responsible for handling I$ and ITLB misses. IMB is a place holder for both the Miss information and the data coming from MS. Let's assume that each instruction is 4 bytes wide, line size is 32bytes, Page size (i.e the size of pages in Main Memory) is 8KB and Main Memory is 32PetaBytes. Lets assume a 64-bit Virtual Address (VA). Since the Cache size is 32KB, the size of each Way will be 4KB (i.e 32KB/8). The branch prediction logic here consists of a 7-bit Gshare register, an array of 2-bit saturating counters to predict branches and a branch target address buffer (BTA) for storing the target address of taken branches (non PC relative branches). An array of 2-bit saturating counters form BPA (Branch Prediction Array) which is organized as a 128x16 structure. BTA is organized as a 128x62 structure. A saturating counter here refers to the following - When the counter reaches its maximum count (i.e 2'b11 in this case) further increments to its count value will not change the value (i.e will remain at 2'b11). Since CALL/RETURN instructions are not being supported (for simplicity reasons), FU does not provide support for a RAS (Return Address Stack) structure. Typically RAS gets used to push PC of the CALL (results in calling a Subroutine) instruction and a RETURN (returns from Subroutine) instruction pops this value back from RAS wherein this popped value plus four gets used for next instruction access. Having a RAS structure greatly improves the performance of CALL/RETURN instructions. Here FU has 2 stages namely **A** and **F** (*Figures 48 and 53 in the following pages provides micro-architectural description for the functionality in pipe stages **A** and **F***). Following are the things that get done in each of the 2 stages - **A Stage -** **1.** Access all 8-Ways of I$ Data as well as Tag array with the lower bits (i.e [11 :5]) of the Virtual Address (since the minimum page size is 8KB, the lower 13 bits of Virtual Address (VA) should be same as the lower 13 bits of Physical Address (PA), so accessing the Cache with bits [11:5] of VA should not cause any aliasing issues). Since the line size is 32bytes we will have 128 lines sitting in each Way. Each Way is organized as a 128x256 array (i.e 128 rows with each row having 8 instructions). On every access we access 8 instructions from each Way as shown in Figure 48. Since each Way has 128 lines we need to store Tag addresses for all 128 lines. Tag array is implemented as a 128 entry structure with each entry holding Tag address (i.e bits [44:12]) for a line stored in its corresponding Data Way. Since the line size is 32bytes, we use bits [11:5] of the VA to access each Way of the Tag array. Since ITLB is fully associative we compare bits [63:13] of the VA against all the entries of VAA (Virtual Address Array). If one of the entries results in a HIT then we end up with a 64-bit one hot vector which gets used to select its corresponding entry

Table 17: Functionality within Various Blocks

Block	Functionality
	holding on to bits [44:13] of the PA in PAA (Physical Address Array). The PA coming out of FA-ITLB along with bit 12 of the VA gets compared against the ones read out of each of the Tag arrays. In the case of a HIT we end up with a 8-bit one hot vector which gets flopped at the end of this stage. The 8 instructions from each Way gets flopped at the end of this stage. **2.** On every access to I$, BPA and BTA also gets accessed. The value that gets used to index BPA is an XORed value of the index bits used to access I$ (i.e VA[11 :5]) and 7-bit Gshare register value. Gshare register which is initially initialized with all 0's is a shift left register. This register is left shifted every time a branch instruction is found to be predicted taken and the shift in value into this register is the least significant bit (i.e bit 2 (i.e [2]) in this case) of the target address of the branch instruction which was predicted as taken. We read a 16-bit value from BPA (i.e [15:14] for I0, [13:12] for I1, [11:10] for I2, [9:8] for I3, [7:6] for I4, [5:4] for I5, [3:2] for I6 and [1:0] for I7) and a 62-bit value (i.e [63:2]) from BTA and flop them at the end of this stage. The most significant bit of the 2-bit value read from BPA for each instruction is the prediction bit and the least significant bit is the strength bit. This 2-bit value can have one of the following four combinations | Prediction Bit	Strength Bit	
---	---	---
0	0	Strongly Not Taken
0	1	Weakly Not Taken
1	0	Weakly Taken
1	1	Strongly Taken	 The 2-bit saturating counters in the BPA are all initialized to 2'b01 at the time of RESET. BPA along with BTA are updated by CU with the result of the branch instruction. BTA gets updated only for non-PC relative branch instructions. In the case of a I$ or ITLB miss the miss information gets forwarded to IMB which is responsible for getting the data and updating I$/ITLB for the missed addresses. **F Stage -** **1.** Select instructions from one of the 8 Ways using 8-bit one hot vector as Way select and forward the selected fetch group to DU/Instruction Buffer. **2.** Instructions are decoded to see if there are any branch instructions and if there are branch instructions then 2-bit Gshare corresponding to the branch instruction gets used to predict whether it needs to be taken or not taken. If its predicted taken then the instructions following the branch instruction are discarded and the new PC and NPC values are calculated and used to access I$. If the branch is predicted not taken then there is no change in the flow of execution. In many cases there is a 1-2 cycle penalty once we come across a branch instruction which is predicted taken. This penalty is for the reason that while we are predicting a branch in F stage there is already an access happening in A stage and if this branch were to be predicted taken then we need to cancel the current access in A stage and make a new I$ access. **3.** Generate and forward 3-bit Valid Vector to DU. This provides information about the validity of instructions in the current fetch group being forwarded to DU. **4.** See if any fetched instruction results in any kind of exception (i.e Parity error etc.) and if it does result in any kind of exception then forward this information to DU along with the instruction.

Table 17: Functionality within Various Blocks

Block	Functionality
DU	It maintains the following logic - **1.** 5-bit counter (since all the WRF's are 32 entries) for renaming destination registers (i.e assigning WRF_ID) of instructions forwarded by FU. Issue Queue and Commit Queue uses this 5-bit WRF_ID (i.e IID) to update its entries with instructions forwarded by DU. **2.** 6-bit counter (since LQ is 32 entries deep, the additional bit (i.e bit [5] of the 6-bit counter value) is the wrap bit that gets used by LQ in determining the age of the instructions sitting in it) for assigning Load Queue ID's (i.e LQ_ID) for Load and Store instructions that gets forwarded by FU. LQ_ID[4:0] gets used by DCU as index to update LQ entries with its corresponding Load/Store instruction. **3.** 5-bit counter (since SQ is 16 entries deep, the additional bit (i.e bit [4] of the 5-bit counter value) is the wrap bit that gets used by SQ in determining the age of the Store instructions sitting in it)) for assigning Store Queue ID's (i.e SQ_ID) for Store instructions forwarded by FU. SQ_ID[3:0] gets used by DCU as index to update SQ entries with its corresponding Store instruction. **4.** Resource management logic for managing various resources in various units down the pipe. DU has 1 pipe stage namely **D** (*Figures 48 and 53,54 in the following pages provides micro-architectural description for the functionality in pipe stage* **D**). Following are the things that get done in **D** stage - **D Stage -** **1.** Generate predecode bits as required by units (i.e RIU, EXU, DCU and CU) down the pipe. **2.** Manage the following resources in various units down the pipeline - 32-entry Issue Queue (in RIU), 32-entry Integer Working Register File (in EXU), 32-entry Floating Point Working Register File (in EXU), 32-entry Condition Code Working Register File (in EXU), 32-entry Load Queue (in DCU), 16-entry Store Queue (in DCU) and 32-entry Commit Queue (in CU) **3.** Provide Instruction ID's (i.e IID) for all the valid instructions that gets forwarded by FU. **4.** Rename destination registers (i.e assigning IWRF_ID, FWRF_ID and CWRF_ID) of all the valid instructions that gets forwarded by FU. **5.** Provide Load Queue ID's (i.e LQ_ID) for all the valid Loads and Stores forwarded by FU. **6.** Provide Store Queue ID's (i.e SQ_ID) for all the valid Stores forwarded by FU. **7.** Provide a 3-bit Slot information (i.e Slot Vector[2:0]) for each valid instruction forwarded by FU. Slot Vector corresponding to an instruction conveys the following information to RIU - a. if Slot Vector[2:0] = 3'b001, this instruction can be issued only to Slot0 in EXU. b. if Slot Vector[2:0] = 3'b010, this instruction can be issued only to Slot1 in EXU. c. if Slot Vector[2:0] = 3'b100, this instruction can be issued only to Slot2 in EXU. Since most of the Queue structures are less than or equal to the size of Commit Queue, we only have to manage Commit Queue and Store Queue. The reason we have to manage Store Queue even though its size is less than CQ is because Store instructions which are committed (i.e drained from CQ) need not necessarily be completed (i.e written to local caches and seen globally by other Processors) yet by the SQ. Also since we are providing WRF_ID to all the valid instructions forwarded by FU, IID = WRF_ID for any valid instruction.

Table 17: Functionality within Various Blocks

Block	Functionality
RIU	It has a 32 entry IRT (Integer Rename Table), 32 entry FRT (Floating point Rename Table) and a 8 entry CRT (Condition Code Rename Table). The reason IRT is 32 entries, FRT is 32 entries and CRT is 8 entries is because the Core provides support for 32 Integer architectural registers, 32 Floating Point architectural registers and 8 Condition Code architectural registers. Architectural registers are the ones seen by the programmer. It has a 32 entry Issue Queue (IQ) where the instructions from DU go and sit before getting picked for issue. It has a 32-entry Ready Bit Array (RBA). RBA is indexed with the renamed source register specifier (i.e WRF_ID). If the entry in RBA indexed by WRF_ID has 1 in it then it means that the source register is dependent free. Here for simplicity its been assumed that floating point (fp) and graphics instructions only use fp registers and fp&g condition codes where as integer instructions only use integer registers and integer condition codes (except for Load and Store instructions). IRT, FRT and CRT are implemented as register file structures for area, timing and power reasons.
	IRT has 12 read ports (9 read ports to support source register renaming of 3 integer instructions (each of which can have a maximum of 3 integer source registers) forwarded by D stage, 3 read ports to support reading of data for invalidation based on commit information provided by CU) and 6 write ports (3 write ports to update entries with the new renamed destination register specifiers of 3 integer instructions (each of which can have one integer destination register) provided by D stage, 3 write ports to invalidate entries based on commit information provided by CU). Here Read happens in the first half and Write happens in the second half of the cycle.
	FRT has 12 read ports (9 read ports to support source register renaming of 3 floating point (fp) and graphics instructions (each of which can have a maximum of 3 fp source registers) forwarded by D stage, 3 read ports to support reading of data for invalidation based on commit information provided by CU) and 6 write ports (3 write ports to update entries with the new renamed destination register specifiers of 3 floating point and graphics instructions (each of which can have one fp destination register) provided by D stage, 3 write ports to invalidate entries based on commit information provided by CU). Here Read happens in the first half and Write happens in the second half of the cycle.
	CRT has 6 read ports (3 read ports to support source register renaming of 3 condition code (CC) integer and floating point and graphics (fp&g) instructions (each of which can have a maximum of one condition code source register) forwarded by D stage, 3 read ports to support reading of data for invalidation based on commit information provided by CU) and 6 write ports (3 write ports to update entries with the new renamed destination integer and fp&g condition code register specifiers of 3 integer/fp&g instructions (each of which can have one integer or fp&g destination CC register) provided by D stage, 3 write ports to invalidate entries based on commit information provided by CU). Here Read happens in the first half and Write happens in the second half of the cycle.
	Here we could have optimized logic by having a separate Valid bit array for each of the Rename tables but to make things simpler for explanation I have combined it with each of the rename tables.
	Since EXU is 3-Way Superscalar (i.e it has 3 Slots with each Slot having one or more functional units), RIU provides support for 3 pickers (i.e one for each Slot (i.e Slot0, Slot1 and

Table 17: Functionality within Various Blocks

Block	Functionality
	Slot2)) each of which can pick an instruction Out-Of-Order (i.e a younger instruction could be picked for issue ahead of an older instruction). RIU has 3 pipe stages namely **R, UI** and **P** (*Figures 49, 50 and 54, 55, 56 in the following pages provides micro-architectural description for the functionality in pipe stages* **R**, **UI** *and* **P**). Following are the things that get done in each of the three stages - **R Stage** - **1.** Rename source registers by doing intra-bundle dependency checking and by looking at the rename tables using architectural register specifiers (5-bit register specifiers in the instruction opcode) as index. **2.** Update rename tables with the new values forwarded by DU i.e index rename tables with valid destination architectural register specifiers forwarded by DU and update those entries with renamed destination register values (i.e WRF_ID's) provided by DU. **3.** Invalidate entries in rename tables based on commit data provided by CU. It invalidates an entry if the following condition is true <u>Condition</u> WRF_ID of the retiring instruction (i.e instruction being committed by CU) matches with the WRF_ID sitting in the rename table at location indexed by the architectural destination register specifier (i.e ARF_ID) of the retiring instruction. **4.** Provide 'use_WRF_data' (if set to Logic 1 means use data from WRF) bit for each of the source register fields (i.e RS1, RS2 and RS3) in an instruction. 'use_WRF_data' bit for a source register field say 'RS1' is set if one of the following 2 conditions is true <u>Condition 1</u> a. Instruction has valid 'RS1' field. b. Register field 'RS1' matches with one of the destination register field (i.e 'RD') of an instruction older to it in the same fetch group. <u>Condition 2</u> a. Instruction has valid 'RS1' field. b. Rename table indexed with 'RS1' as index has a valid entry in it. c. ARF_ID of the committing instruction does not match with 'RS1' or if it matches the WRF_ID sitting in the indexed entry (i.e indexed by ARF_ID of the committing instruction) does not match with the WRF_ID of the committing instruction. **5.** Provide 'Ready' bit (if set to Logic 1 means source register is dependent free) for each of the source register fields (i.e RS1, RS2 and RS3) in an instruction. 'Ready' bit for a source register field say 'RS1' is set if one of the following conditions is true <u>Condition 1</u> Source register specifier 'RS1' is invalid (i.e the instruction does not have a valid 'RS1' field). <u>Condition 2</u> Valid bit in its corresponding entry in the rename table shows the entry to be invalid and 'RS1' does not have a match with any of the architectural destination register specifiers of older valid instructions in its fetch group (i.e fetch group forwarded by D stage). <u>Condition 3</u> Valid bit in its corresponding entry in the rename table shows the entry to be valid, 'RS1' does not have a match with any of the architectural destination register specifiers of older valid instructions in its fetch group (i.e fetch group forwarded by D stage) and ARF_ID (architectural destination register specifier) of the retiring instruction matches with RS1 and the WRF_ID of the retiring instruction matches with the WRF_ID sitting in the rename table indexed by the ARF_ID of the retiring instruction.

Table 17: Functionality within Various Blocks

Block	Functionality
	UI Stage - **1.** Re-evaluate ready status of the incoming instructions from R stage by looking at the Ready Bit Array (RBA). 'Ready' bit for a source register specifier is set if one of the following conditions is true *Condition 1* 'Ready' bit for the source register specifier forwarded by R stage shows logic 1(i.e ready) in it. *Condition 2* 'Ready' bit for the source register specifier forwarded by R stage shows logic 0 (i.e not ready) in it but indexing RBA with renamed source register specifier (i.e WRF_ID of the source register specifier) as index shows 1 in its corresponding entry. **2.** Set entries in RBA with data from P stage i.e destination WRF_ID's of the picked instructions in P stage gets used to index RBA to set its corresponding entry to 1. Destination WRF_ID's of the committing instructions index RBA and resets its corresponding entries. **3.** Re-evaluate 'use_WRF_data' bit for each of the source register fields by comparing the WRF_ID's of the committing instructions against the renamed source register fields of the instructions forwarded by R stage. If there is a match then the 'use_WRF_data' bit for the matching renamed source register field is reset to 0 and if there is no match then the bit retains its value from R stage. **4.** Update IQ with these new instructions from R stage. **P Stage -** **1.** Re-evaluate 'Ready' bit status for each of the renamed source register specifiers of the instructions sitting in the Issue Queue by comparing the renamed source register specifiers (i.e WRF_ID's) of the instructions against the renamed destination register specifiers of the instructions picked for issue in the previous cycle. The bit gets set to Logic 1 if there is a match and remains unmodified otherwise. **2.** Re-evaluate 'use_WRF_data' bit for each of the renamed source register specifiers of the instructions sitting in the Issue Queue by comparing the source register specifiers of the instructions against the WRF_ID of the retiring instructions. The bit gets reset to Logic 0 if there is a match and remains unmodified otherwise. **3.** Each picker picks an instruction for issue based on the slot vector associated with the instruction, ready status of the instruction and the age of the instruction. For an instruction to be picked by Slot0 picker the following three conditions need to be true - a. Slot Vector = 3'b001. b. Instruction has no register dependency (i.e all its source registers are dependent free). c. Instruction is the oldest among the instructions sitting in the IQ which are dependent free and have their slot vector equal to 3'b001. Similarly for Slot1 picker and Slot2 picker. **4.** Provide the necessary controls to the ARF, WRF and bypass Mux's. **5.** Update Issue Queue with the updated 'Ready' bit status and 'use_WRF_data' bit status for each of the source register specifiers of the instruction.

Table 17: Functionality within Various Blocks

Block	Functionality
EXU	EXU has three Slots (i.e Slot0, Slot1 and Slot2) to support superscalar nature of the pipe. Slot0 has two functional units - Integer ALU and Branch execution unit (BEU). Integer ALU is responsible for handling all the integer arithmetic, logic and condition code instructions whereas BEU is responsible for handling all the branch instructions. Type of instructions that get executed in this Slot are - *(see instruction format table below)* Slot1 has 3 functional units - Integer ALU, Integer Multiplier and Integer Divider. Integer ALU is responsible for handling all the integer arithmetic and logic instructions along with all the Load/Store instructions for calculating Virtual Address. Integer Multiplier is responsible for handling all the integer multiply instructions and integer Divider is responsible for handling all the integer divide instructions. Type of instructions that get executed in this Slot are -

Instruction	Assuming the following Instruction Format and Operation
ADD R1, R2, R3	31 - - - - - - - - - - - - 0 | Opcode | R1 | R2 | R3 | R1 + R2 -> R3, add the contents of R1 with the contents of R2 and write the result into R3
SUB R1, R2, R3	31 - - - - - - - - - - - - 0 | Opcode | R1 | R2 | R3 | R1 - R2 -> R3, substract the contents of R2 from the contents of R1 and write the result into R3
ADDcc R1, R2, R3	31 - - - - - - - - - - - - 0 | Opcode | R1 | R2 | R3 | R1 + R2 -> R3, CC, add the contents of R1 with the contents of R2 and write the result into R3 and modify condition codes
ADDC R1, R2, R3	31 - - - - - - - - - - - - 0 | Opcode | R1 | R2 | R3 | R1 + R2 + C -> R3, add the contents of R1, R2 and carry field of the condition code and write the result into R3
MOVcc R1, R2, R3	31 - - - - - - - - - - - - 0 | Opcode | R1 | R2 | R3 | Move based on condition code the contents of either R1 or R2 into R3
ADD R1, Imm_Data, R3	31 - - - - - - - - - - - - 0 | Opcode | R1 | Imm_Data | R3 | Add the contents of R1 with Imm_Data and write the result into R3
MOVR R1, R2, R3	31 - - - - - - - - - - - - 0 | Opcode | R1 | R2 | R3 | Move based on R3 the contents of either R2 or R3 into R3
BRcc	31 - - - - - - - - - - - - 0 | Opcode | Imm_Data | Branch based on condition code

Table 17: Functionality within Various Blocks

Block	Functionality

Instruction	Assuming the following Instruction Format and Operation
ADD R1, R2, R3	 `| Opcode	R1	R2	R3 |` R1 + R2 -> R3, add the contents of R1 with the contents of R2 and write the result into R3
LD R1, R2, R3	 `| Opcode	R1	R2	R3 |` Load the value from memory location ([R1]+[R2]) into R3
LD R1, R2, FR1	 `| Opcode	R1	R2	FR1 |` Load the value from memory location ([R1]+[R2]) into FR1
ST R1, R2, R3	 `| Opcode	R1	R2	R3 |` Store the value in R3 into memory location pointed by ([R1]+[R2])
ST R1, R2, FR1	 `| Opcode	R1	R2	FR1 |` Store the value in FR1 into memory location pointed by ([R1]+[R2])
ADD R1, Imm_Data, R3	 `| Opcode	R1	Imm_Data	R3 |` Add the contents of R1 with Imm_Data and write the result into R3
MOVR R1, R2, R3	 `| Opcode	R1	R2	R3 |` Move based on R1 the contents of either R2 or R3 into R3
MUL R1, R2, R3	 `| Opcode	R1	R2	R3 |` Multiply the contents of R1 with the contents of R2 and write the result into R3
DIV R1, R2, R3	 `| Opcode	R1	R2	R3 |` Divide the contents of R1 with the contents of R2 and write the result into R3

Slot2 has 3 functional units - Floating point and Graphics (FPG) ALU, FPG Multiplier and FPG Divider. FPG ALU is responsible for handling all the floating point and graphics arithmetic, logic and condition code instructions. FPG Multiplier is responsible for handling all the fp and graphics multiply instructions. FPG Divider is responsible for handling all the fp and graphics divide instructions. Type of instructions that get executed in this Slot are -

Table 17: Functionality within Various Blocks

Block	Functionality

Instruction	Assuming the following Instruction Format and Operation
FADD FR1, FR2, FR3	31 ─ ─ ─ ─ ─ ─ ─ ─ ─ ─ ─ 0 Opcode | FR1 | FR2 | FR3 FR1 + FR2 -> FR3, add the contents of FR1 with the contents of FR2 and write the result into FR3
FADDcc FR1, FR2, FR3	31 ─ ─ ─ ─ ─ ─ ─ ─ ─ ─ ─ 0 Opcode | FR1 | FR2 | FR3 FR1 + FR2 -> FR3, CC, add the contents of FR1 with the contents of FR2 and write the result into FR3 and modify condition codes
FADDC FR1, FR2, FR3	31 ─ ─ ─ ─ ─ ─ ─ ─ ─ ─ ─ 0 Opcode | FR1 | FR2 | FR3 FR1 + FR2 + C -> FR3, add the contents of FR1, FR2 and carry field of the condition code and write the result into FR3
FMOVcc FR1, FR2, FR3	31 ─ ─ ─ ─ ─ ─ ─ ─ ─ ─ ─ 0 Opcode | FR1 | FR2 | FR3 Move based on condition code the contents of either FR1 or FR2 into FR3
FADD FR1, Imm_Data, FR2	31 ─ ─ ─ ─ ─ ─ ─ ─ ─ ─ ─ 0 Opcode | FR1 | Imm_Data | FR2 Add the contents of FR1 with Imm_Data and write the result into FR2
FMUL FR1, FR2, FR3	31 ─ ─ ─ ─ ─ ─ ─ ─ ─ ─ ─ 0 Opcode | FR1 | FR2 | FR3 Multiply the contents of FR1 with the contents of FR2 and write the result into FR3
FDIV FR1, FR2, FR3	31 ─ ─ ─ ─ ─ ─ ─ ─ ─ ─ ─ 0 Opcode | FR1 | FR2 | FR3 Divide the contents of FR1 with the contents of FR2 and write the result into FR3

It has a 32 entry IWRF, 32 entry FWRF, 32 entry CWRF, 32 entry IARF, 32 entry FARF and a 8 entry CARF. Here IWRF, FWRF, CWRF, IARF, FARF and CARF are all implemented as register file structures.

IWRF has 9 read ports (3 read ports to support instructions with three integer source registers issued to Slot0, 3 read ports to support instructions with three integer source registers issued to Slot1, 3 read ports to read data from IWRF to be written into IARF or to be forwarded to DCU (for integer Stores) as we can commit three integer instructions in a given cycle) and 3 write ports (1 write port for Slot0, 1 write port for Slot1, 1 write port for data from DCU (i.e for integer Loads)).

FWRF has 5 read ports (2 read ports to support floating point and graphics instructions with two floating point source registers issued to Slot2, 3 read ports to read data from FWRF to be written into FARF or to be forwarded to DCU (for floating point Stores) as we can commit three floating point and graphics instructions in a given cycle) and 2 write ports (1 write port for Slot2, 1 write port for data from DCU (i.e for floating point Loads)).

CWRF has 5 read ports (1 read port to support instructions with integer condition code source register issued to Slot0, 1 read port to support instructions with fp & g condition code source register issued to Slot2, 3 read ports to read data from CWRF to be written into CARF as we can commit three condition code modifying instructions in a given cycle) and 2 write ports (1 write port for Slot0, 1 write port for Slot2).

IARF has 6 read ports (3 read ports to support instructions with three integer source registers issued to Slot0, 3 read ports to support instructions with three integer source registers issued to Slot1) and 3 write ports (3 write ports to update IARF with data from IWRF as we can commit three integer instructions in a given cycle).

Table 17: Functionality within Various Blocks

Block	Functionality
	FARF has 2 read ports (2 read ports to support floating point and graphics instructions with two floating point source registers issued to Slot2) and 3 write ports (3 write ports to update FARF with data from FWRF as we can commit three floating point and graphics instructions in a given cycle). CARF has 2 read ports (1 read port to support instructions with integer condition code source register issued to Slot0, 1 read port to support instructions with fp & g condition code source register issued to Slot2) and 3 write ports (3 write ports to update CARF with data from CWRF as we can commit three condition code (CC) modifying instructions in a given cycle). EXU has 4 pipestages namely **RD, E, UW** and **W** *(Figures 51, 52 and 56, 57 in the following pages provides micro-architectural description for the functionality in pipe stages* **RD, E, UW** *and* **W** *)*. Following are the things that get done in each of the four stages - **RD Stage -** **1.** Here instructions picked for issue read their data from either IWRF, IARF, FWRF, FARF, CWRF or CARF based on the validity of their register specifiers, type of register specifier (i.e integer, floating point or condition code) and read from (i.e from WRF or ARF) information provided by P stage. The read data goes through final bypass Mux which has any one of the following inputs as one of its legs based on the Slot it gets issued to - data from IWRF, data from IARF, data from FWRF, data from FARF, data from CWRF, data from CARF, data from Slot0, data from Slot1, data from Slot2, Immediate data forwarded by P stage and data from DCU. **E Stage -** **1.** Each Slot executes instructions based on the operands and control information received from RD stage. If Slot1 gets an 'ADD' instruction then ALU1 is active while the rest of the functional units within this Slot (i.e Multiplier and Divider) are inactive. **2.** Slot0 generates a 64-bit integer value and a 8-bit integer condition code value which gets flopped at the end of this stage and also gets bypassed to the bypass Muxes in RD stage. **3.** Slot1 generates a 64-bit integer value which gets flopped at the end of this stage and also gets bypassed to the Muxes in RD stage. **4.** Slot2 generates a 64-bit floating point and graphics value and a 8-bit fp & g condition code value which gets flopped at the end of this stage and also gets bypassed to the Muxes in RD stage. **5.** Completion and exception reports for all the instructions being executed are generated and flopped at the end of this stage. **UW Stage -** **1.** Data values (i.e 64-bit integer data and 8-bit integer CC data from Slot0, 64-bit integer data from Slot1, 64-bit floating point and graphics data and 8-bit fp & g CC data from Slot2) forwarded by E stage are written into the appropriate WRF's (i.e IWRF, FWRF or CWRF) **2.** Forward completion and exception report of instructions that got executed in E stage to CU. **W Stage -** **1.** Move data from WRF's to ARF's or DCU based on control information from CU.

Table 17: Functionality within Various Blocks

Block	Functionality
DCU	It has a 32 entry Load Queue (LQ), 16 entry Store Queue (SQ), 32KB, 8-Way set associative D\$ (1 Read, 1 Write Port Data Array, 2 Read, 2 Write Port Tag Array), a 64 entry fully associative DTLB and a 128x4 Way set associative MMU (Memory Management Unit). Read/Write port in Data array and 1 Read/Write port in Tag array are shared between Load and Store requests. The additional Read/Write port in Tag array is used by MS for invalidations (i.e for the case of external snoop requests). D\$ here is Virtually indexed and Physically tagged. Let's assume the line size to be 32bytes. Since the Cache size is 32KB, the size of each Way will be 4KB (i.e 32KB/8). D\$ here is non-blocking (i.e a Load miss does not block other Loads from being processed) and write through (i.e Store hit updates both D\$ as well as MS). Loads are allocating (i.e in the case of a Load miss the data corresponding to the Load coming from MS gets written into D\$) while Stores are non-allocating (i.e in the case of a Store miss, Stores do not update D\$ with its data). Control logic here provides support for 1, 2, 4 and 8 byte Loads and Stores with alignment and zero or sign extension for Loads. Also here its assumed that the Core supports TSO (Total Store Order) memory ordering where all the older Loads complete before a younger Load completes, all the older Loads completes before a younger Store completes and all the older Stores complete (i.e is committed and is seen by all other Processors within the System) before a younger Store completes. MMU here provides address translation for the case where you miss in DTLB or ITLB. LQ uses LQ_ID forwarded by EXU (which gets it from RIU) to update its entries with the Loads and Stores forwarded by EXU. SQ uses SQ_ID forwarded by EXU to update its entries with the Stores forwarded by EXU. LQ and SQ uses the most significant bit of their LQ_ID and SQ_ID (i.e LQ_ID[5] and SQ_ID[4]) forwarded by EXU to determine the age of Loads and Stores sitting in their respective entries. Some of the fields in a LQ entry are VA/PA, LQ_ID[5:0]/SQ_ID[4:0], Valid bit etc. Some of the fields in a SQ entry are PA, SQ_ID[4:0], Valid bit, Data[63:0] etc. A younger Load could be processed by the LQ before an older Load gets processed (this can happen as Loads get issued out-of-order by the Issue Queue). Eventhough Loads can be processed out-of-order by the LQ, they are freed in program order. Freeing of the Load (i.e say Load1) in the LQ does not necessarily mean that Load (Load1 here) gets retired by CU. While a Load gets processed by the LQ, it is still snoopable until it gets freed by the LQ. If a match is detected (i.e a completed but not freed Load address matches with the Snoop address) then LQ forces CU to re-ifetch all the instructions younger to the Load which had a match along with the matching Load instruction. A Load issued from the LQ accesses DTLB along with D\$. Each entry in LQ has Virtual Address forwarded by EXU in E stage. DTLB provides address translation (i.e VA to PA) for the Load instruction which gets forwarded to both the D\$ and LQ. D\$ provides the data for the Load in case of a Hit. In case of a Miss the Load request gets forwarded to MS through some arbitration logic. The inputs to the arbitration logic listed in priority are - 1. I\$ miss request from FU 2. Load request from LQ 3. Instruction prefetch request from FU 4. Data prefetch request from DCU 5. Store request from SQ. LQ compares its Loads against all the Stores sitting in the SQ to see if there is a RAW (Read after Write) hit. For a RAW hit to happen the following needs to be true - Physical Address of the Load matches with the Physical Address of an older Store sitting in the SQ. In such case we wait for the Store to complete (i.e drained from SQ) and then get data for the Load which had a RAW

Table 17: Functionality within Various Blocks

Block	Functionality
	hit from either D$ or MS (if the Store had a Miss in D$). Here we are not bypassing data (i.e for the Load) from SQ to make things simpler and doing this also saves power as we have got rid of all the bypass logic which would have otherwise been needed.

A Store issued from the LQ accesses DTLB along with D$. DTLB provides address translation for the Store instruction which gets forwarded to both the D$ and SQ. D$ provides the Hit/Miss information for Stores and this information goes and sits in SQ. Stores are completed in-order from the SQ i.e an younger Store is completed only after all the older Stores are completed.

A Load which hits in D$ forwards the completion and exception report to CU once it gets the Hit information. In the case of a miss it forwards the completion and exception report when it receives data for the Load from MS. A Store sends the completion and exception report once it gets Hit/Miss information from D$.

As Loads and Stores get drained from LQ and SQ once they have completed execution, the drain count of these Loads and Stores gets forwarded to DU so that it can properly manage LQ and SQ.

DCU has 4 pipe stages namely **AT, DA, DT** and **UW** *(Figure 46 below provides microarchitectural description for the functionality in pipe stages* **AT, DA** *and* **DT***)*. Following are the things that get done in each of the four stages -

AT Stage -
1. Virtual Address along with the rest of the control signals for the Load/Store forwarded by EXU is flopped at the end of this stage.

DA Stage -
1. Here we access all 8-Ways of D$ Data as well as Tag array with the lower bits (i.e [11:3]) of the Virtual Address (since the minimum page size is 8KB, the lower 13 bits of Virtual Address (VA) should be same as the lower 13 bits of Physical Address (PA), so accessing the Cache with bits [11:3] of VA should not cause any aliasing issues). Since the line size is 32bytes we will have 128 lines sitting in each Way. Each Way is organized as a 128x256 array (i.e 128 rows with each row having 32bytes of data). On every access we access 64-bit data from each Way as shown in Figure 46 below. Since each Way has 128 lines we need to store 128 Tag addresses for all the 128 lines. Tag array is implemented as a 128 entry structure with each entry holding Tag address (i.e bits [44:12]) for a line stored in its corresponding Data Way. Since the line size is 32bytes, we use bits [11:5] of VA to access each Way of Tag array. Since DTLB is fully associative we compare bits [63:13] of VA against all the entries of VAA (Virtual Address Array). If one of the entries results in a HIT then we end up with a 64-bit one hot vector which gets used to select its corresponding entry holding on to bits [44:13] of the PA in PAA (Physical Address Array). The PA coming out of FA-DTLB along with bit 12 of the VA gets compared against the ones read out of each of the Tag arrays. In the case of a HIT we end up with a 8-bit one hot vector which gets flopped at the end of this stage.
2. Update LQ and SQ with the incoming Load and Store instructions with LQ_ID and SQ_ID as index. |

Table 17: Functionality within Various Blocks

Block	Functionality
	DT Stage - 1. Select 64-bit data from one of the 8 Ways using 8-bit one hot vector as Way select and forward the aligned and zero or sign extended data to EXU in the case of Loads. 2. See if there is a RAW hit by comparing the physical address of the Load against older Stores sitting in the SQ. If there is a match then prevent the Load from forwarding data to EXU but rather wait for the matching Store to complete (i.e the Store is drained from SQ and the data has been updated in both Level 1 (i.e if there is a Hit in Level 1 Cache) and Level 2 Caches). Once the Store has completed, retry Load again by reading the data from either Level 1 or Level 2 Caches and forward it to EXU after formatting and zero or sign extending it. 3. Determine Parity error for the Load and if there is one then force a Miss in Level 1 Cache (i.e D$) and get the data from Level 2 (i.e L2$) before forwarding it to EXU. **UW Stage -** 1. Forward completion and exception report for the Load/Store to CU. Figure 46: Functionality in Pipe Stage AT, DA and DT

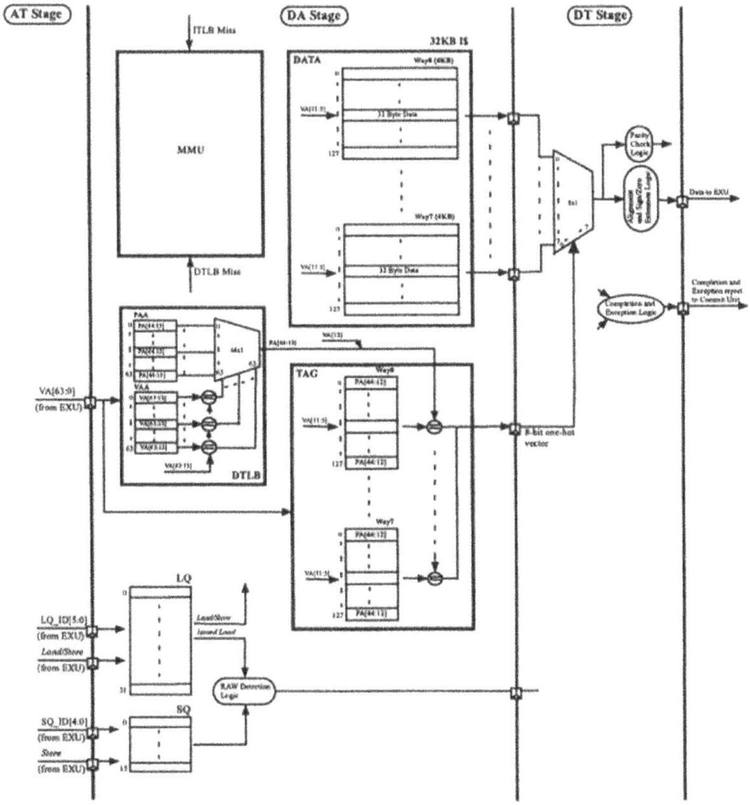

Table 17: Functionality within Various Blocks

Block	Functionality
CU	It has a 32 entry Commit Queue (CQ). CU uses instruction ID (IID) forwarded by various units to index into CQ. It can commit any combination of 3 instructions in a given cycle. All the instructions are committed in program order i.e a younger instruction cannot be committed until all the instructions older to it have been committed. A requirement for any instruction to be ready for commit is it has executed without an exception and it is the oldest among the group of instructions sitting in the CQ. A 'DIV' instruction could result in a 'div_by_zero' exception if EXU finds that the divisor for the DIV instruction is zero. A Load Word(i.e load 4bytes of data) operation could result in a 'mem_address_not_aligned' exception if DCU finds that the least significant 2-bits of the VA are non zero for the Load. Whenever CU comes across an instruction which results in an exception then it does not commit this instruction but rather does the following things - 1. Flush the pipe (i.e invalidate all the instructions currently in the pipe) and force all the various counters in various units to initialize to its RESET value. 2. Invalidate all the entries in CQ or reset the pointers used to access CQ. 3. Send new PC and NPC values to FU. These new values will result in loading the software trap handler code to take care of the exception. CU also sends a count of the number of instructions it is committing in a given cycle to DU so that DU can properly manage the CQ resource. CU has 2 pipe stages namely **UW** and **W**. Following are the things that get done in each of the two stages - **UW Stage -** 1. Update CQ with data sent from various units using IID as index. 2. Pick 3 instructions for commit. 3. Send retire count to DU. 4. Send necessary control information to RIU to update its Rename Tables, Ready Bit Array and Issue Queue. **W Stage -** 1. Forward necessary control information to EXU for data transfer from desired WRF's to ARF's.
MS	It has a 4MB, 8 Way, 1 Read port, 1 Write port L2$, 2 Memory Controllers (MC0, MC1), Cross Bar Interface, Memory Subsystem Controller and System Interface Unit. L2$ here supports MHOSI Cache Coherency Protocol as this CMP Chip is assumed to be designed for Multiprocessor systems. Since L2$ is 4MB, the size of each Way will be 512KB (i.e 4MB/8). Let's assume the line size to be 64Bytes. L2$ here is physically organized as 4 banks with each bank having all 8 Ways as shown in Figure 47 below. L2$ is banked mainly to improve bandwidth and save power while not all banks are being accessed. L2$ is inclusive to reduce the amount of snoop traffic going into Level 1 Caches of each of the individual Cores. L2$ here is responsible for the following things 1. Handle requests from Cross Bar Interface. 2. Provide data in the case of a Hit and place requests to one of the memory controllers (i.e MC0 or MC1) or System Interface Unit in the case of a Cache Miss. 3. Maintain Cache Coherency. 4. Provide and maintain ECC protection for Data and Parity protection for Tag. 5. Provide support for BIST and BISI.

Table 17: Functionality within Various Blocks

Block	Functionality
	Memory controllers (i.e MC0 and MC1) here are mainly responsible for handling requests that missed in L2$. Cross Bar Interface here is responsible for handling requests from Core0 thru CoreN. System Interface Unit here is responsible for handling all the snoop traffic from the system bus and the requests from Memory Subsystem Controller. Memory Subsystem Controller here manages inflow and outflow of data to/from L2$, Cross Bar Interface, Memory Controllers and SIU.

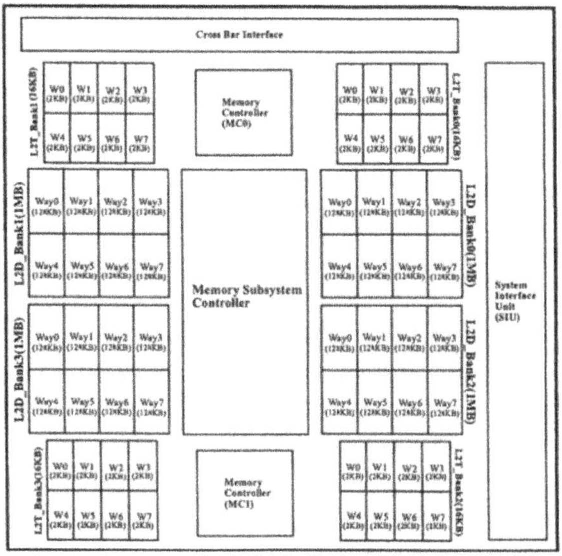

Figure 47: Memory Subsystem

Figures 48 thru 57 below show how 3 integer instructions (I0 (ADD R1, R2, R9), I1 (SUB R4, R5, R11) and I2 (ADD R9, R11, R13) where, I2 younger than I1 younger than I0) belonging to a fetch group get executed in various pipe stages as they flow from **A** stage all the way to **W** stage.

Table 18 below gives a brief description of the interface between various blocks within the Core.

Table 18: Interface between Various Blocks within the Core

↗	FU	DU	RIU	EXU	DCU	CU
FU	-	Forward fetch group (i.e a max of three instructions)	-	-	I$ miss, ITLB miss and Instruction Prefetch requests	-
DU	Stall request	-	Forward desired predecode information (i.e slot ID, instruction type, register valid specifiers etc.), renamed destination specifiers, instruction ID, load queue ID and store queue ID along with the instructions	-	-	Forward desired predecode information (i.e instruction type, register valid specifiers etc.), renamed and architectural destination register specifiers and exception report for each instruction
RIU	-	-	-	Forward necessary control information to access the desired entries in various WRF's and ARF's, bypass MUX's and functional units	Forward load queue ID and store queue ID along with some desired predecode information for Loads and Stores that gets issued	-
EXU	-	-	-	-	Forward virtual address for Loads/Stores along with data to be stored for Stores	Forward completion and exception report for each instruction
DCU	Forward requested instruction line (i.e for I$ miss, Instruction prefetch) or address translation (i.e for ITLB miss)	Forward a count of the number of load queue and store queue entries being drained/released	D$ miss information	Data for Loads	-	Forward completion and exception report for each Load/Store instruction
CU	Forward pipe flush information and new PC and NPC values in the case of branch misprediction, exception, interrupt or RESET	Forward a count of the number of commit queue (CQ) entries being drained/released along with pipe flush and throttle information	Forward instruction commit information (i.e architectural destination register specifier, renamed destination register specifier etc.) along with pipe flush information	Forward necessary control information (i.e architectural destination register specifier, renamed destination register specifier etc.) for transfer of data from desired WRF's to desired ARF's	Commit information for Stores so that store queue in DCU can mark the Store as being committed. Forward pipe flush information	-

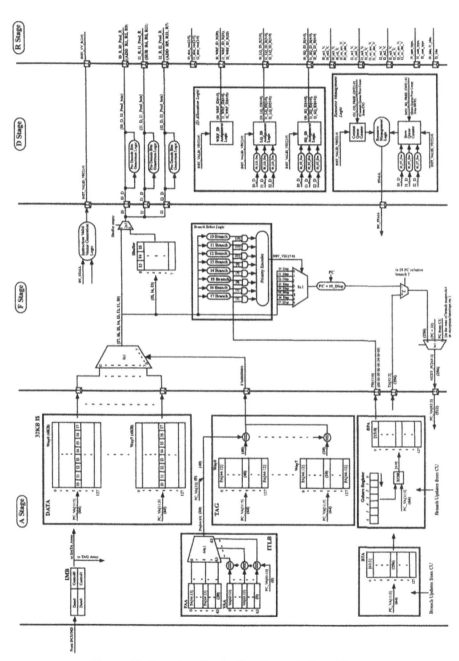

Figure 48: Functionality in Pipe Stage A, F and D

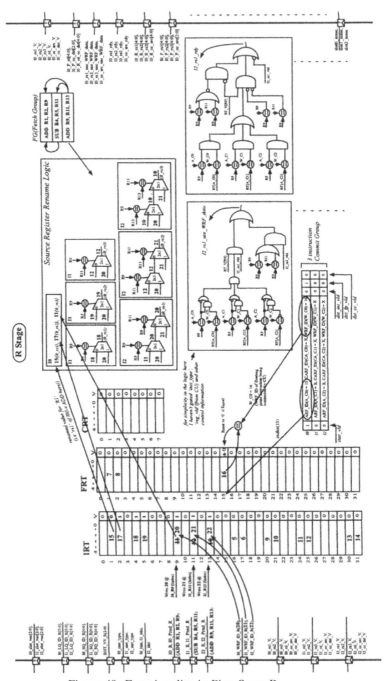

Figure 49: Functionality in Pipe Stage R

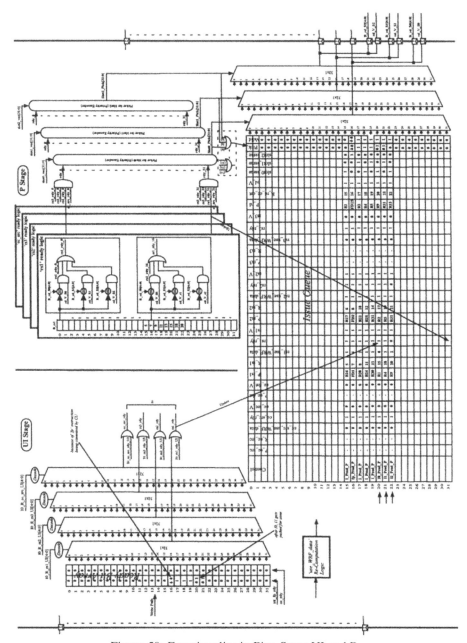

Figure 50: Functionality in Pipe Stage UI and P

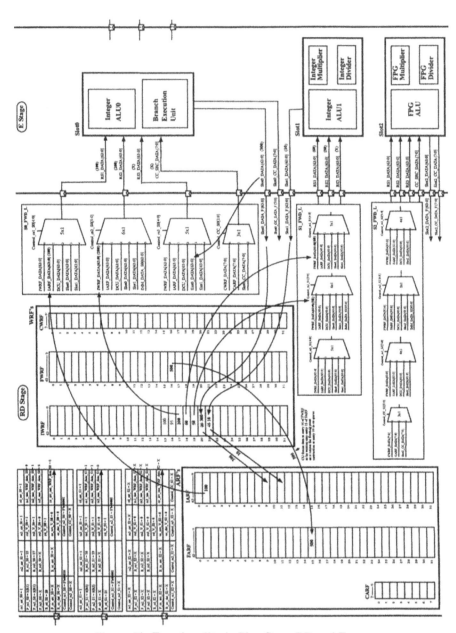

Figure 51: Functionality in Pipe Stage RD and E

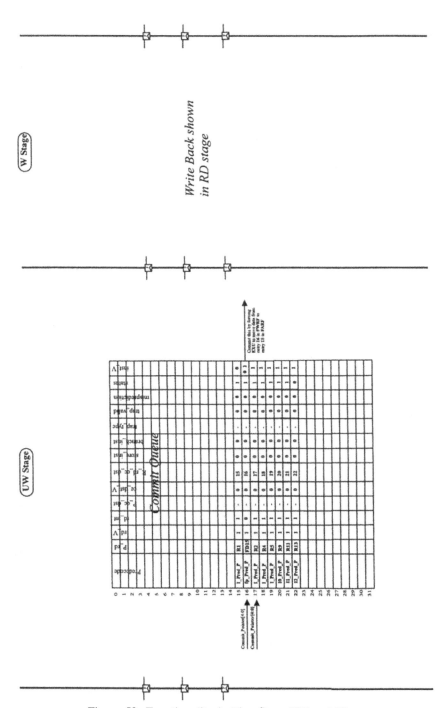

Figure 52: Functionality in Pipe Stage UW and W

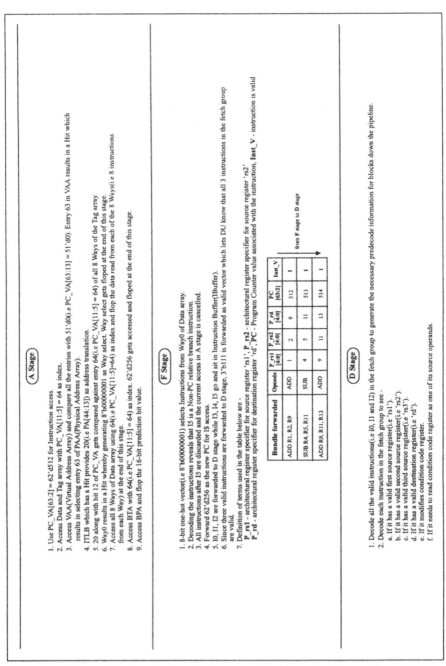

A Stage

1. Use PC_VA[63:2] = 62'd512 for Instruction access.
2. Access Data and Tag array with PC_VA[11:5] = 64 as index.
3. Access VAA(Virtual Address Array) and compare all the entries with 51'd0(i.e PC_VA[63:13] = 51'd0). Entry 63 in VAA results in a Hit which results in selecting entry 63 of PAA(Physical Address Array).
4. ITLB which has a Hit provides 20(i.e PA[44:13]) as address translation.
5. 20 along with bit 12 of PC_VA gets compared against entry 64(i.e PC_VA[11:5] = 64) of all 8 Ways of the Tag array.
6. Way0 results in a Hit whereby generating 8'b00000001 as Way select. Way select gets floped at the end of this stage.
7. Access all 8 Ways of Data array using 64(i.e PC_VA[11:5]=64) as index and flop the data read from each of the 8 Ways(i.e 8 instructions from each Way) at the end of this stage.
8. Access BTA with 64(i.e PC_VA[11:5] = 64) as index. 62'd256 gets accessed and floped at the end of this stage.
9. Access BPA and flop the 16-bit prediction bit value.

F Stage

1. 8-bit one-hot vector(i.e 8'b00000001) selects Instructions from Way0 of Data array.
2. Decoding the instructions reveals that I5 is a Non-PC relative branch instruction.
3. All instructions after I5 are discarded and the current access in A stage is cancelled.
4. Forward 62'd256 as the new PC for IS access.
5. I0, I1, I2 are forwarded to D stage while I3, I4, I5 go and sit in Instruction Buffer(IBuffer).
6. Since three valid instructions are forwarded to D stage, 3'b111 is forwarded as valid vector which lets DU know that all 3 instructions in the fetch group are valid.
7. Definition of terms used in the table below are -
   P_rs1 - architectural register specifier for source register 'rs1', P_rs2 - architectural register specifier for source register 'rs2'
   P_rd - architectural register specifier for destination register 'rd', PC - Program Counter value associated with the instruction, Inst_V - instruction is valid

Bundle forwarded	Opcode	P_rs1 [4:0]	P_rs2 [4:0]	P_rd [4:0]	PC [63:2]	Inst_V
ADD R1, R2, R9	ADD	1	2	9	512	1
SUB R4, R5, R11	SUB	4	5	11	513	1
ADD R9, R11, R13	ADD	9	11	13	514	1

from F stage to D stage

D Stage

1. Decode all the valid instructions(i.e I0, I1 and I2) in the fetch group to generate the neccessary predecode information for blocks down the pipeline.
2. Decode each instruction in the fetch group to see -
   a. If it has a valid first source register(i.e 'rs1').
   b. If it has a valid second source register(i.e 'rs2').
   c. If it has a valid third source register(i.e 'rs3').
   d. If it has a valid destination register(i.e 'rd').
   e. If it modifies condition code register.
   f. If it needs to read condition code register as one of its source operands.

Figure 53: Description for Functionality shown in Figure 48

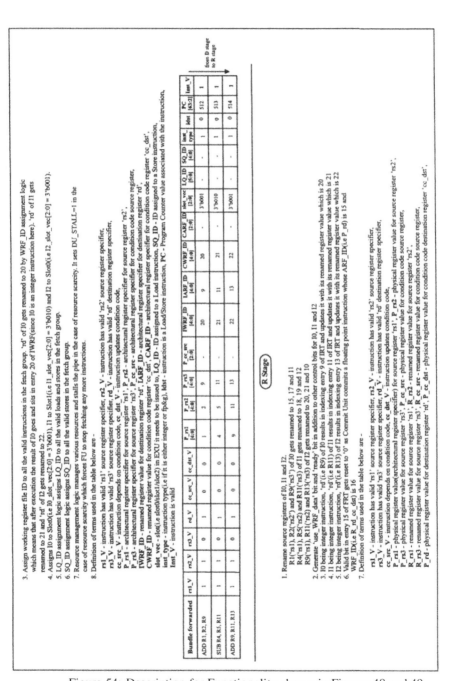

Figure 54: Description for Functionality shown in Figures 48 and 49

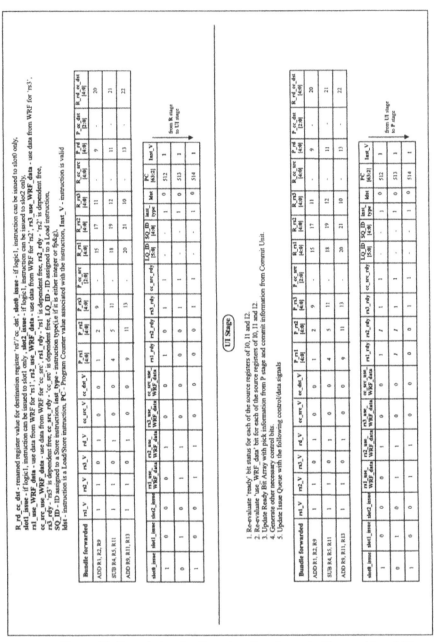

Figure 55: Description for Functionality shown in Figures 49 and 50

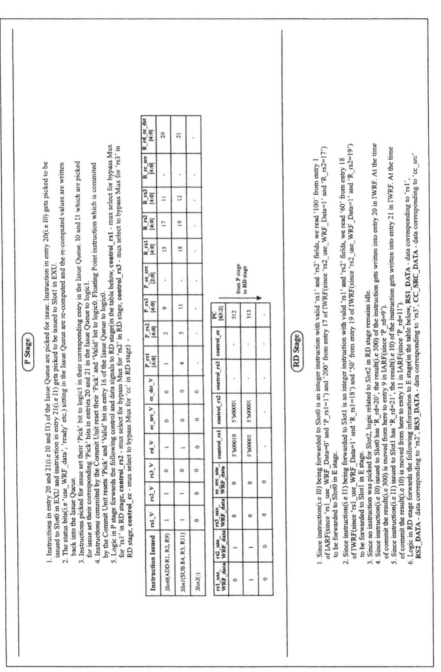

Figure 56: Description for Functionality shown in Figures 50 and 51

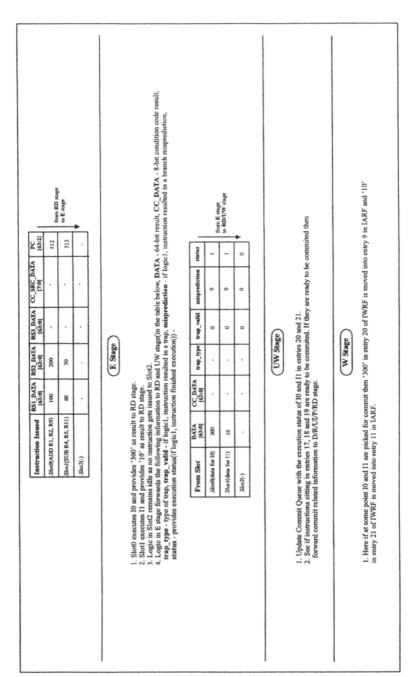

Instruction Issued	RS1_DATA [63:0]	RS2_DATA [63:0]	RS3_DATA [63:0]	CC_SRC_DATA [7:0]	PC [63:2]
Slot0(ADD R1, R2, R9)	100	200	-	-	512
Slot1(SUB R4, R5, R11)	60	50	-	-	513
Slot2(-)	-	-	-	-	-

from RD stage
to E stage

E Stage

1. Slot0 executes I0 and provides '300' as result to RD stage.
2. Slot1 executes I1 and provides '10' as result to RD stage.
3. Logic in Slot2 remains idle as no instruction gets issued to Slot2.
4. Logic in E stage forwards the following information to RD and UW stage(in the table below, DATA - 64-bit result, CC_DATA - 8-bit condition code result,
   trap_type - type of trap, trap_valid - if logic1, instruction resulted in a trap, misprediction - if logic1, instruction resulted in a branch misprediction,
   status - provides execution status(if logic1, instruction finished execution)) -

From Slot	DATA [63:0]	CC_DATA [63:0]	trap_type	trap_valid	misprediction	status
Slot0(data for I0)	300	-	-	0	0	1
Slot1(data for I1)	10	-	-	0	0	1
Slot2(-)	-	-	-	0	0	0

from E stage
to RD/UW stage

UW Stage

1. Update Commit Queue with the execution status of I0 and I1 in entries 20 and 21.
2. See if instructions sitting in entries 17, 18 and 19 are ready to be commited. If they are ready to be commited then
   forward commit related information to D/R/UI/P/RD stage.

W Stage

1. Here if at some point I0 and I1 are picked for commit then '300' in entry 20 of IWRF is moved into entry 9 in IARF and '10'
   in entry 21 of IWRF is moved into entry 11 in IARF.

Figure 57: Description for Functionality shown in Figures 51 and 52

19. What is a Multithreaded Processor and what are the various threading techniques used in a Multithreaded Processor?

A Processor supporting multiple threads (a thread is a schedulable software entity; It is equivalent to a Solaris lightweight process) is known as a Multithreaded Processor. Two terms most commonly associated with Multithreading are Horizontal waste and Vertical waste. *Horizontal waste* occurs when some but not all of the execution slots can be used. *Vertical waste* occurs when a execution cycle goes completely unused. Multithreading helps in reducing Horizontal waste and Vertical waste. Figure below shows Horizontal and Vertical waste in a 2-Way Superscalar Processor supporting single Thread.

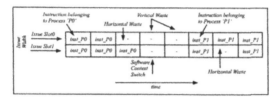

Figure 58: Horizontal Waste and Vertical Waste

Table below shows the various threading techniques used in a Multithreaded Processor.

Table 19: Threading Techniques used in a Multithreaded Processor

Threading Technique	Description
Vertical Threading (VT)	A coarse grain processor scheduling technique in which instructions from one particular thread occupies a given pipe stage. Multiple threads share superscalar processing resources in aggregate but not in the same cycle. The motivation here is to simplify the scheduling of execution timeslots. A typical VT switching algorithm as applied to an out-of-order, 2-way superscalar processor supporting two threads is shown in Figure 59 below. It shows a 5 stage frontend pipeline (with switching logic in D stage) for such a processor.

Table 19: Threading Techniques used in a Multithreaded Processor

Threading Technique	Description
	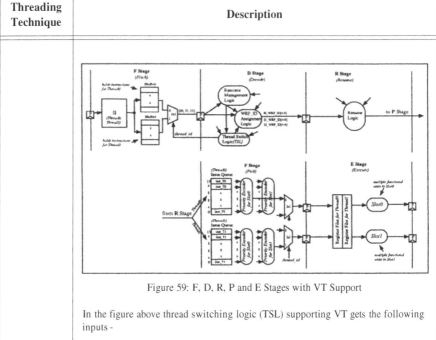

Figure 59: F, D, R, P and E Stages with VT Support

In the figure above thread switching logic (TSL) supporting VT gets the following inputs -

1. Decode unit (in D stage) which maintains an active thread (thread which is currently active in the pipe) counter (which gets initialized to zero while there is a thread switch and is incremented every cycle thereafter) forwards the active thread count value to TSL *(TSL looks at active thread count value mainly to avoid thread starvation; here the maximum count value the active thread count gets compared against could be set by the software or could be a hardwired value).*
2. Decode unit (in D stage) which manages all the various resources down the pipe (Load Queue, Store Queue, Commit Queue etc.) for each thread provides resource scarcity information to TSL.
3. Fetch unit (in F stage) which manages IBuffers for both the threads provides buffer empty information to TSL.
4. Commit unit down the pipe (assuming an out of order processor) provides the flush pipe (while there is a branch mispredict or exception or interrupt etc.) information for each thread to TSL.

In the figure above, TSL in D stage switches threads based on the flowchart shown below.

Table 19: Threading Techniques used in a Multithreaded Processor

Threading Technique	Description
	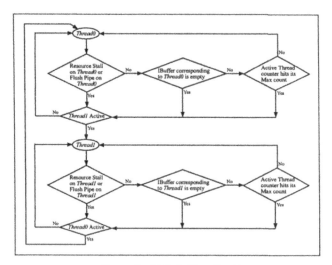 Figure 60: VT Switching Algorithm Figure 61 below shows instruction issue as a function of time for a single threaded, out-of-order, 2-way superscalar processor and a 2 threaded, out-of-order superscalar processor supporting VT. 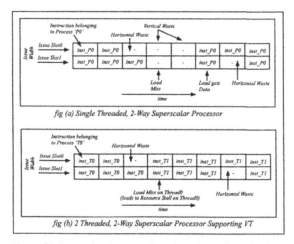 Figure 61: Instruction Issue as a Function of Time for a Single Threaded Processor and a 2 Threaded Processor Supporting VT Figure above illustrates how a processor supporting VT results in reducing Vertical

Table 19: Threading Techniques used in a Multithreaded Processor

Threading Technique	Description
	waste when compared to a single threaded, 2-way superscalar processor.
Simultaneous Multithreading (SMT)	A fine grain process or scheduling technique that permits multiple independent threads to issue instructions to a superscalar's functional units in a single cycle. SMT combines the multiple-instruction-issue features of wide superscalar processors with the latency-hiding ability of multithreaded architectures. On an SMT processor, all hardware threads are active simultaneously, competing each cycle for all available resources. This dynamic sharing of processor resources enables SMT to exploit thread-level and instruction-level parallelism interchangeably. Here both forms of parallelism can be effectively used to increase processor utilization. A typical SMT switching algorithm as applied to an out-of-order, 2-way superscalar processor supporting two threads is shown in Figure 62 below. It shows a 5 stage pipeline (with switching logic in D stage) for such a processor. 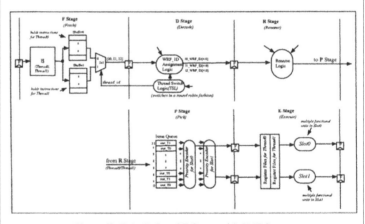 Figure 62: F, D, R, P and E Stages with SMT Support

Table 19: Threading Techniques used in a Multithreaded Processor

Threading Technique	Description
	In the figure above, TSL in D stage switches threads based on the flowchart shown in figure below. 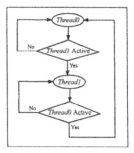 Figure 63: SMT Switching Algorithm Figure 64 below shows instruction issue as a function of time for a single threaded, out-of-order, 2-way superscalar processor and a 2 threaded, out-of-order superscalar processor supporting SMT. 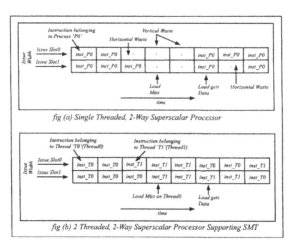 Figure 64: Instruction Issue as a Function of Time for a Single Threaded Processor and a 2 Threaded Processor Supporting SMT From figure above we see that by feeding fetch groups from either threads in a round robin fashion and allowing instructions from both threads to participate in resource (functional units) sharing (shown in P and E stage) we have avoided both horizontal and vertical waste.

Table 19: Threading Techniques used in a Multithreaded Processor

Threading Technique	Description
Branch Threading (BT)	A coarse grain processor scheduling technique where the thread switch is based on hitting a branch instruction in a particular thread. The motivation here is to avoid branch penalty where the processor doesn't provide support for static or dynamic branch prediction techniques. A typical BT switching algorithm as applied to an out-of-order, 2-way superscalar processor supporting two threads is shown in Figure 65 below. It shows a 5 stage frontend pipeline (with switching logic in D stage) for such a processor. 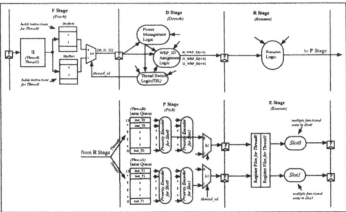 Figure 65: F, D, R, P and E Stages with BT Support In the figure above thread switching logic (TSL) supporting BT gets the following inputs - 1. Decode unit (in D stage) which maintains an active thread (thread which is currently active in the pipe) counter (which gets initialized to zero while there is a thread switch and is incremented every cycle thereafter) forwards the active thread count value to TSL *(TSL looks at active thread count value mainly to avoid thread starvation; here the maximum count value the active thread count gets compared against could be set by the software or could be a hardwired value).* 2. Decode unit (in D stage) which decodes instructions in the fetch group to see if there are any branches forwards this information to TSL. 3. Fetch unit (in F stage) which manages IBuffers for both the threads provides buffer empty information to TSL. 4. Commit unit down the pipe (assuming an out of order processor) provides the flush pipe (while there is a branch mispredict or exception or interrupt etc.) information for each thread to TSL. In the figure above, TSL in D stage switches threads based on the flowchart shown below.

Table 19: Threading Techniques used in a Multithreaded Processor

Threading Technique	Description
	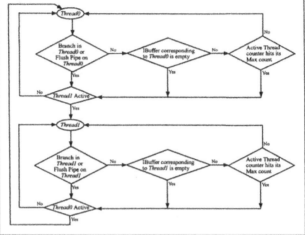 Figure 66: BT Switching Algorithm Figure 67 below shows instruction issue as a function of time for a single threaded, out-of-order, 2-way superscalar processor with no static or dynamic branch prediction and a 2 threaded, out-of-order superscalar processor supporting BT with no static or dynamic branch prediction. 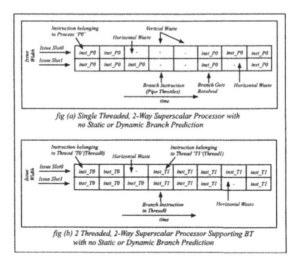 Figure 67: Instruction Issue as a Function of Time for a Single Threaded Processor and a 2 Threaded Processor Supporting BT

Table 19: Threading Techniques used in a Multithreaded Processor

Threading Technique	Description
	Figure above illustrates how BT results in reducing Vertical waste when compared to a single threaded, 2-way superscalar processor which throttles on branches.
Power Threading (PT)	A coarse grain processor scheduling technique where the thread switch is based on power dissipated by a particular thread being processed. The motivation here is to keep the average power dissipated within Spec. A typical PT switching algorithm as applied to an out-of-order, 2-way superscalar processor supporting two threads is shown in Figure 68 below. It shows a 5 stage frontend pipeline (with switching logic in D stage) for such a processor. 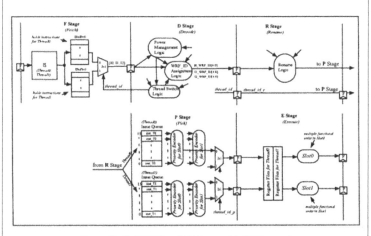 Figure 68: F, D, R, P and E Stages with PT Support In the figure above thread switching logic (TSL) supporting PT gets the following inputs - 1. Decode unit (in D stage) which maintains an active thread(thread which is currently active in the pipe) counter (which gets initialized to zero while there is a thread switch and is incremented every cycle thereafter) forwards the active thread count value to TSL *(TSL looks at active thread count value mainly to avoid thread starvation; here the maximum count value the active thread count gets compared against could be set by the software or could be a hardwired value).* 2. Decode unit (in D stage) which manages power dissipated for each thread provides this information to TSL *(here the high threshold value for power dissipation could be set by the software or could be a hardwired value).* 3. Fetch unit (in F stage) which manages IBuffers for both the threads provides buffer empty information to TSL. 4. Commit unit down the pipe (assuming an out of order processor) provides the flush pipe (while there is a branch mispredict or exception or interrupt etc.) information for each thread to TSL.

Table 19: Threading Techniques used in a Multithreaded Processor

Threading Technique	Description
	In Figure 68 above, TSL in D stage switches threads based on the flowchart shown below. Figure 69: PT Switching Algorithm

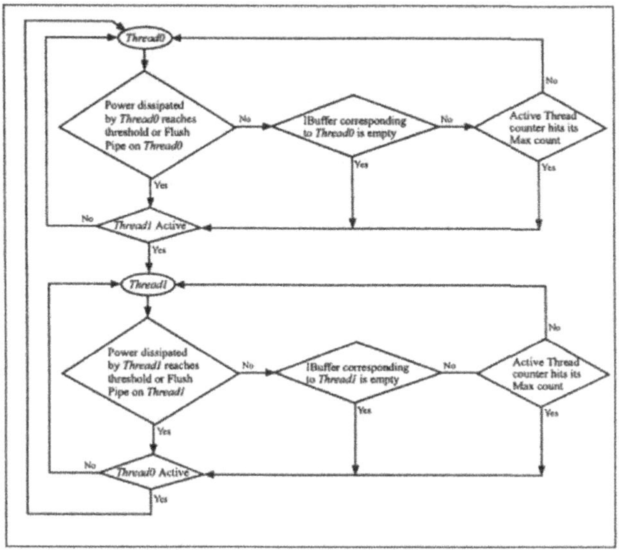

20. What are the most common Data Protection Schemes used for Caches?

Table below shows the most common Data Protection Schemes used for Caches.

Table 20: Data Protection Schemes for Caches

Scheme	Description
Providing Odd Parity	This scheme requires following things to happen - 1. Generate Odd Parity bit(s) (the bit is set to 1 if the count of number of 1's in the data bits it is associated with is even and is set to 0 if the count of the number of 1's in the data bits it is associated with is odd) for the Data that needs to be written into the Cache. The number of Parity bits generated really depends on the number of Data bits we want to cover with one Parity bit. If we want to have one Parity bit per 8-bit data word then we need to generate 8 Parity bits for a 64-bit Data word. 2. Update the Cache with the Data along with the Odd Parity bit(s).

Table 20: Data Protection Schemes for Caches

Scheme	Description
	3. When the Data is read (here you read both Data as well as the Odd Parity bit(s) associated with it) regenerate the Odd Parity bit(s) and compare it against the Parity bit(s) read from the Cache. If they match then the Data is Valid otherwise it is bad (i.e corrupted). Figure below shows the Architectural Path for Odd Parity generation (Data here is shown as a 32-bit wide word with one Parity bit across the entire word), Equation for Odd Parity generation and an Example showing Odd Parity as applied to a 8-bit data word. If the Data is bad then the architecture normally calls for one of the following things to happen 1. Take a Software Trap (i.e here the Software Trap handler replaces the bad Data with the good one from either higher level Caches or Main Memory) or 2. Force a Miss for the request and reload the Data from either higher level Caches or Main Memory. If the Cache under question is a write-back Cache and the data under question is in 'Modified' state (i.e this is the only one which has the latest updated data) then the architecture handles it differently *(in many cases it takes a RESET trap where the trap handler Reset's the CPU)*.

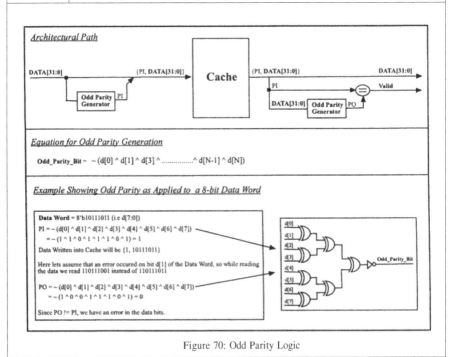

Figure 70: Odd Parity Logic

Table 20: Data Protection Schemes for Caches

Scheme	Description
Providing Even Parity	This scheme requires following things to happen - **1.** Generate Even Parity bit(s) (the bit is set to 1 if the count of number of 1's in the data bits it is associated with is odd and is set to 0 if the count of the number of 1's in the data bits it is associated with is even) for the Data that needs to be written into the Cache. The number of Parity bits generated really depends on the number of Data bits we want to cover with one Parity bit. If we want to have one Parity bit per 8-bit data word then we need to generate 8 Parity bits for a 64-bit Data word. **2.** Update the Cache with the Data along with the Even Parity bit(s). **3.** When the Data is read (here you read both Data as well as the Even Parity bit(s) associated with it) regenerate the Even Parity bit(s) and compare it against the Parity bit(s) read from the Cache. If they match then the Data is Valid otherwise it is bad (i.e corrupted). Figure below shows the Architectural Path for Even Parity generation (Data here is shown as a 32-bit wide word with one Parity bit across the entire word), Equation for Even Parity generation and an Example showing Even Parity as applied to a 8-bit data word. If the Data is bad then the architecture normally calls for one of the following things to happen 1. Take a Software Trap (i.e here the Software Trap handler replaces the bad Data with the good one from either higher level Caches or Main Memory) or 2. Force a Miss for the request and reload the Data from either higher level Caches or Main Memory. If the Cache under question is a write-back Cache and the data under question is in 'Modified' state (i.e this is the only one which has the latest updated data) then the architecture handles it differently *(in many cases it takes a RESET trap where the trap handler Reset's the CPU).*

Table 20: Data Protection Schemes for Caches

Scheme	Description

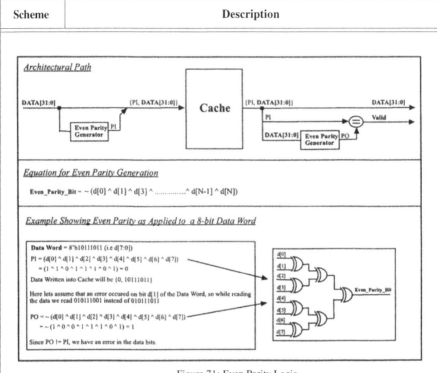

Figure 71: Even Parity Logic

Providing ECC through Hamming Code	This scheme requires following things to happen - **1**. Generate *Partial Parity* bits and *Whole Parity* bit for the Data that needs to be written into the Cache. **2**. Form a *Code Word* with the *Partial Parity* bits, *Whole Parity* bit and Data bits. **3**. Update the Cache with the *Code Word.* **4**. When the Data is read (i.e *Code Word*) generate *Check* bits and *Universal Parity* bit to see if the Data is Valid or Corrupted. Figure below shows the Architectural Path for ECC as applied to a Data Word (Data here is shown as a 32-bit wide word), Equations associated with ECC and an Example showing ECC as applied to a 4-bit data word. Following paragraphs give a more detailed explanation of the terms associated with Hamming Code. Hamming Code requires the generation of more than one parity bit for the correction of single bit error and the detection of double bit error. The parity bits generated here are labeled as *Partial Parity* (PP) bits and *Whole Parity* (WP) bit. There are multiple

Table 20: Data Protection Schemes for Caches

Scheme	Description
	Partial Parity bits and one *Whole Parity* bit. *Partial Parity* bits get used in the correction of single bit error and occupy power of two positions in the *Code Word* whereas *Whole Parity* bit gets used in the detection of double bit error and occupies the most significant bit position in the *Code Word*. *Code Word* here is the word formed by the concatenation of parity and data bits with *Partial Parity* bits occupying the power of two bit positions, *Whole Parity* bit occupying the most significant bit position and the data bits occupying the remaining positions.
	Each *Partial Parity* bit here is assigned to a group of data bits. Data bits associated with a particular PP bit really depends on the position of the PP bit in the *Code Word*. The First PP bit (i.e in location 1 (i.e 2^0) of the *Code Word*) is assigned to data bits in positions 3, 5, 7, 9, 11, 13, ...(i.e positions which have 1 in their 2^0 bit position), Second PP bit (i.e in location 2 (i.e 2^1) of the *Code Word*) is assigned to data bits in positions 3, 6, 7, 10, 11, 14, 15, ...(i.e positions which have 1 in their 2^1 bit position), Third PP bit (i.e in location 4 (i.e 2^2) of the *Code Word)* is assigned to data bits 5, 6, 7, 12, 13, 14, 15, ...(i.e positions which have 1 in their 2^2 bit position) etc. Each of the PP bits maintain even parity across the data bits it is associated with (i.e it is set to 0 if the count of the number of 1's in the data bits it is associated with is even and is set to 1 if the count of the number of 1's in the data bits it is associated with is odd).
	The *Whole Parity* bit is assigned to all the data bits and PP bits. It maintains even parity across all the data bits and PP bits (i.e it is set to 0 if the count of the number of 1's in the bits it is associated with is even and is set to 1 if the count of the number of 1's in the bits it is associated with is odd).
	To find if there is a single bit error or double bit error, *Check* bits and *Universal Parity* (UP) bit are generated. All the *Check* bits are concatenated to form a Check Bit Vector which gets used to determine if their is a single bit error and if their is one which data bit has an error. The number of *Check* bits generated (in other words the size of the Check Bit Vector) really depends on the number of PP bits associated with the data bits. The least significant bit of the Check Bit Vector (i.e C[0]) is generated by calculating even parity across the first PP bit in position 1 and data bits in positions 3, 5, 7, 9, 11, 13, ... of the read Code Word, C[1] is generated by calculating even parity across the second PP bit in position 2 and data bits in positions 3, 6, 7, 10, 11, 14, 15, ... of the read *Code Word,* C[2] is generated by calculating even parity across the third PP bit in position 4 and data bits in positions 5, 6, 7, 12, 13, 14, 15, ... of the read *Code Word* etc. The *Universal Parity* bit is generated by calculating even parity across the read *Code Word.* The kind of error (i.e single bit or double bit) and position of the error (i.e in the case of single bit error) is determined from the table shown in Figure below.
	If it is found that the Data has double bit error then the architecture normally calls for one of the following things to happen 1. Take a Software Trap (i.e here the Software Trap handler replaces the bad Data with the good one from either higher level Caches or Main Memory) or 2. Force a Miss for the request and reload the Data from either higher level Caches or Main Memory. If the Cache under question is a write-back Cache and the data under question is in 'Modified' state (i.e this is the only one which has the latest updated data) then the architecture handles it differently *(in many cases it takes a RESET trap where the trap handler Reset's the CPU).*

Table 20: Data Protection Schemes for Caches

Scheme	Description
	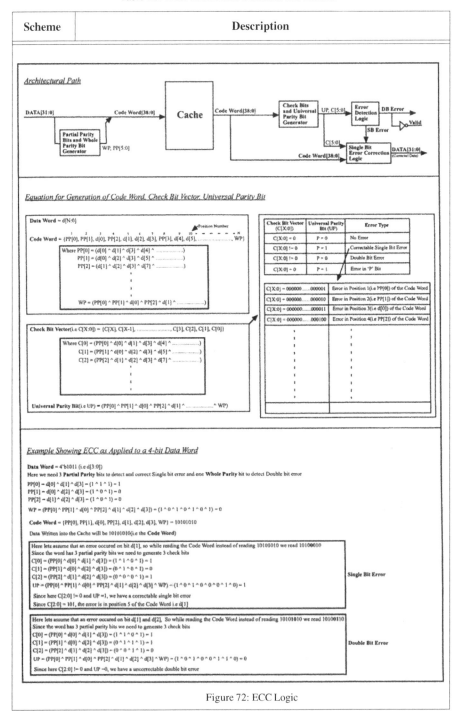

Figure 72: ECC Logic

21. Describe a technique for improving performance (single thread as well as multi thread performance) of a CMP (Chip Multi Processing) chip by sharing Working Register File(s) between adjacent Processor Cores?

The motivation behind this question is to make the reader think that there could be many resources that could be shared between Cores in a CMP chip. Even though people might argue that as we go into deep submicron technologies transistors come for free, but there will be cases where sharing some critical resources between Cores does make sense from performance point of view.

In an out-of-order processor instructions get renamed so that they could be issued out-of-order. In such case an younger instruction which is dependent free could be issued for execution before an older instruction gets issued. Instructions here get renamed to one of the entries in Working Register File (WRF). WRF (in some cases it could be called Working Data Structure/Re-Order Buffer) here refers to the structure which holds on to the result of an instruction immediately after its execution. Generally an out-of-order processor has a WRF, an Issue Queue and a Commit Queue in addition to other Units. Typically an instruction which gets renamed goes and sits in an Issue Queue before it gets picked for execution. While in Issue Queue if the instruction is dependent free and there are no other older instructions which are ready to be issued then it gets picked for execution. After execution the result of the instruction goes and sits in WRF. Once the instruction becomes top of the commit queue entry, Commit Queue looks at the instruction to see if it is ready for retirement by looking at its exception and status reports. If the instruction has not resulted in any kind of exception then it reads the value corresponding to the instruction sitting in WRF and writes it into the desired architectural register file.

In an out-of-order processor with a N-entry Commit Queue, N-entry Issue Queue and a N/2-entry WRF, the performance of such a processor could be improved if the size of WRF be increased to N entries. This could be achieved in a CMP chip by sharing WRF between two neighboring processor cores. Sharing is done when one of the processor cores is dead or, alive but parked. 'Processor core is dead' here means that at the time of fab it was found that the core is bad (i.e one of the reasons could be one of the RAM structures in the core has defects etc.). To support sharing of WRF's, a fuse is provided within each core which gets blown at the time of fab when its found that a particular core is bad. An unblown fuse within a core indicates the core being alive. A core could be Parked by the Operating System (OS) when it finds that it doesn't have enough threads to keep the core busy. A Parked core here means the core is idle doing nothing. In a four core CMP chip as shown in Figure 73 (here its assumed that each CORE has a 32-entry, 6 read/3write port WRF structure) we share WRF's between CORE0 and CORE1, and CORE2 and CORE3. Here it is really important to physically place the WRF blocks within individual cores near the edge in the floorplan (as seen from Figure 73). This will reduce most of the routing congestion between Cores and will also ease WRF access timing. Flowchart in Figure 74 shows the conditions under which CORE0 shares WRF of CORE1. Figure 75 shows the logic involved in sharing WRF's between CORE0 and CORE1. Here its been assumed that any instruction can have a maximum of two source registers and one destination register.

In Figures 74 and 75, I have assumed individual cores to be supporting single strand but it could be extended to cores supporting multiple strands with few updates which are not shown here. Also here when CORE0 shares WRF of CORE1, the read data corresponding to WRF_ID's=1xxxxx comes from WRF of CORE1 and write data corresponding to WRF_ID's=1xxxxx gets written into WRF of CORE1.

Most of the signal names in Figure 75 are self explanatory. Some of the signal name definitions are -

CORE0_ALIVE_AND_UNPARKED	Core0 is alive and unparked
I0_DST_WRF_ID[5:0]	Renamed destination ID for instruction I0 assigned by WRF_ID assignment logic in CORE0
I0_we	Write enable for instruction I0 in CORE0
I0_SRC0_WRF_ID[4:0]	Renamed first source register value for instruction I0 in CORE0
I0_re0	Read enable for first source register(i.e SRC0) of instruction I0 in CORE0
CORE1_I0_DST_WRF_ID[5:0]	Renamed destination ID for instruction I0 assigned by WRF_ID assignment logic in CORE1
CORE1_I0_we	Write enable for instruction I0 in CORE1
CORE1_I0_SRC0_WRF_ID[4:0]	Renamed first source register value for instruction I0 in CORE1
CORE1_I0_re0	Read enable for first source register(i.e SRC0) of instruction I0 in CORE1
CORE1_ALIVE_AND_UNPARKED	Core1 is alive and unparked

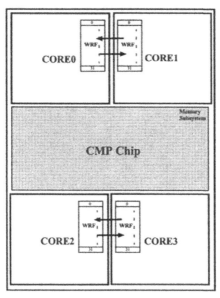

Figure 73: CMP Chip Showing Shared WRF's between Adjacent Cores

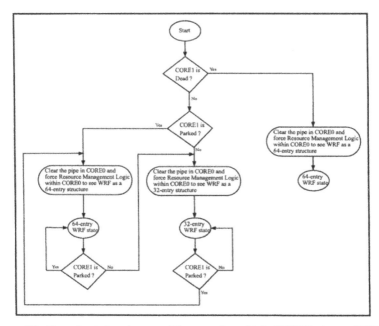

Figure 74: Flowchart showing conditions under which CORE0 shares WRF of CORE1

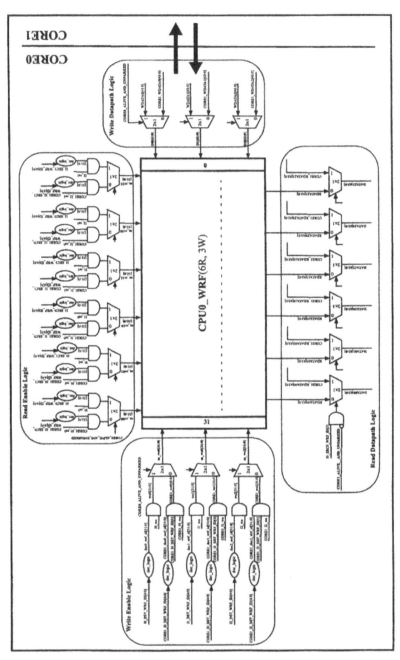

Figure 75: Logic involved in sharing WRF's between CORE0 and CORE1

22. What are some of the techniques used at the Architectural/Microarchitectural level to reduce Power in Caches.

Table below shows some of the techniques used at the Architectural and Microarchitectural level to reduce Power in Caches.

Table 21: Architectural/Microarchitectural Techniques Used to Reduce Power in Caches

1	Going with a Direct Mapped Cache implementation instead of Fully Associative (FA) or Set Associative (SA) Cache implementation where ever possible as FA and SA Caches result in more power dissipation because of the additional logic associated with it (i.e Way select logic, Comparator logic etc.).
2	In a Set Associative Cache, generate your Way select before accessing your Cache (i.e Data portion of it). Doing this will result in accessing only one Way thereby saving power.
3	In the case of a Cache which is banked, selectively generating Read and Write enables for the individual banks of Tag and Data arrays will result in reduced power dissipation.
4	Making use of NOP's in the instruction stream to reduce power in the ICache read access path. To illustrate this lets assume the following ICache implementation - 16KB, 4-Way set associative Cache implemented as 4 banks with *bank0* and *bank1* having *Way0/Way1* and *bank2* and *bank3* having *Way2/Way3* as shown in figure below. Each bank is arranged as a 256x128 structure (i.e it has 256 rows with each row having 4 instructions each (assuming each instruction is 32-bits wide)). Here lets assume the instructions to be interleaved as they get written. In the current illustration a row in *bank0* is arranged as follows - Bit Cell Bit Cell (I0_w0[0] I2_w0[0] I0_w1[0] I2_w1[0]) ---------- (I0_w0[31] I2_w0[31] I0_w1[31] I2_w1[31]) We see from above that each bit cell has 4-bits, 2 corresponding to *way0* and 2 corresponding to *way1*. In a similar fashion a row in *bank0* is arranged as follows - Bit Cell Bit Cell (I1_w0[0] I3_w0[0] I1_w1[0] I3_w1[0]) ---------- (I1_w0[31] I3_w0[31] I1_w1[31] I3_w1[31]) Let's assume a 16 byte line fill to ICache. Here whenever we have line fill from L2$ we look to see if there are any NOP instructions in word position 2 and 3 (as shown in figure below) of the 16 byte line fill (assuming 16 byte line fill is arranged as {word0 (I0), word1 (I1), word2 (I2), word3 (I3)}. If there are NOP's in either word position 2 or 3 then we replace NOP in word position 2 with instruction in position 0 and NOP in word position 3 with instruction in position 1 and we append a predecode bit which gets set for NOP's which got replaced in word positions 2 and 3. This bit tells various units handling this instruction to treat it as a NOP. In figure below we see that we have replaced the first NOP with ADD and the second NOP with SUB and the predecode bit (P) is set for the replaced instructions. This arrangement helps in reducing power in the read access path when I2 and I3 gets accessed from I$. This technique not only results in reducing power but also reduces the effect of coupling and pattern sensitive faults in ICache.

Table 21: Architectural/Microarchitectural Techniques Used to Reduce Power in Caches

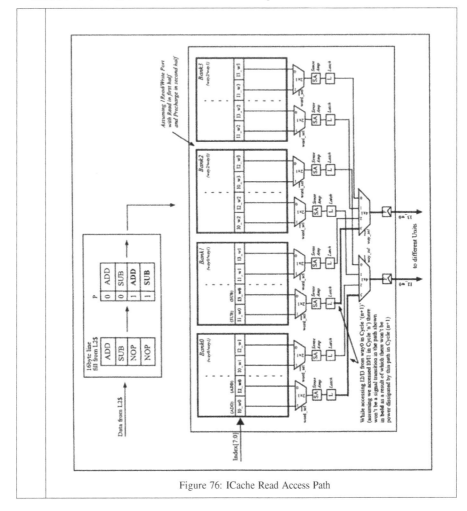

Figure 76: ICache Read Access Path

24. What are the most common benchmarks used to measure CPU/System Performance?

There are currently two roles that benchmarks play in today's world. The first is for customers, as one of several key differentiating points when making a decision and for Organizations which don't have the technical and financial resources to perform benchmark tests can use them as a way of screening out handful of vendors. Ideally, benchmarks provide a standard and meaningful way of measuring a computer system's performance. The second role that benchmarks play is as a marketing tool for vendors. All benchmarks are unique as each one tests an area a little different than the other tests, and some are better measures of a system than others.

Table below shows the most common benchmarks used to measure CPU/System performance.

Table 22: Benchmarks Used to Measure CPU/System Performance

Benchmark	Description
SPEC - Standard Performance Evaluation Corporation is a non-profit corporation formed to establish, maintain and endorse a standardized set of relevant benchmarks that can be applied to the newest generation of high-performance computers	
SPECint2000	Measures compute intensive Integer performance. It contains 12 applications (11 written in C and 1 written in C++ *(Data compression, FPGA circuit placement and routing, C compiler, Minimum cost network flow solver, Chess program, Natural language processing, Ray tracing, Perl, Computational group theory, Object Oriented Database, Data compression utility, Place and route simulator))* that are used as benchmarks. It is measured as a geometric mean of twelve normalized ratios (one for each integer benchmark mentioned above) when compiled with aggressive optimization for each benchmark.
SPECfp2000	Measures compute intensive Floating-Point performance. It contains 14 applications (6 written in Fortran-77, 4 written in Fortran-90 and 4 written in C *(Quantum chromodynamics, Shallow water modeling, Multi-grid solver in 3D potential field, Parabolic/elliptic partial differential equations, 3D Graphics library, Fluid dynamics, Neural network simulation, Finite element simulation, Computer vision, Computational chemistry, Number theory, Finite element crash simulation, Particle accelerator model, application which solves problems regarding temperature, wind, velocity and distribution of pollutants))* that are used as benchmarks. It is measured as a geometric mean of fourteen normalized ratios (one for each floating point benchmark mentioned above) when compiled with aggressive optimization for each benchmark.
SPECintRate	The geometric mean of twelve (uses the same 12 applications as used for SPECint2000) normalized throughput ratios when compiled with aggressive optimization for each benchmark.
SPECfpRate	The geometric mean of fourteen (uses the same 14 applications as used for SPECfp2000) normalized throughput ratios when compiled with aggressive optimization for each benchmark.
SPECjAppServer2002	Measures the performance of a multi-tier workload with the application tier implemented in Java (J2EE) server.
SPECjbb2000	Measures the performance of a Java implemented application tier.
SPEC SFS	Measures NFS file servers performance in terms of throughput and response time.

Table 22: Benchmarks Used to Measure CPU/System Performance

Benchmark	Description
SPEC WEB99	Measures the performance of a web tier application.
TPC - Transaction Processing Performance Council is a non-profit corporation founded to define transaction processing and database benchmarks and to disseminate objective, verifiable TPC performance data to the industry.	
TPC-C	It is a on-line transaction processing (OLTP) benchmark which measures five light-weight transactions against nine tables in a simple workload. The performance metric reported by TPC-C is the number transactions per minute (tpmC).
TPC-H	It is a decision support benchmark which consists of a suite of business oriented ad-hoc queries and concurrent data modifications. The queries and the data populating the database have been chosen to have broad industry-wide relevance. This benchmark illustrates decision support systems that examine large volumes of data, execute queries with a high degree of complexity, and give answers to critical business questions. The performance metric reported by TPC-H is called the TPC-H Composite Query-per-Hour Performance Metric, and reflects multiple aspects of the capability of the system to process queries.
TPC-W	It is a transactional web benchmark. The workload is performed in a controlled internet commerce environment that simulates the activities of a business oriented transactional web server. It simulates three different profiles by varying the ratio of browse to buy: primarily shopping, browsing and web-based ordering. The performance metric reported by TPC-W is the number of web interactions processed per second.
TPC-R	It is a decision support benchmark similar to TPC-H, but here the DBMS optimizers are allowed huge freedom in their use of software technologies. It consists of a suite of business oriented queries and concurrent data modifications. The performance metric reported by TPC-R is called the TPC-R Composite Query-per-Hour Performance Metric, and reflects multiple aspects of the capability of the system to process queries.

2 Logic

1. What are some of the responsibilities of a Logic Designer in the Chip Industry?

Some of the responsibilities of a Logic Designer are summarized below.

Table 23: Logic Designer Responsibilities

1	Come up with a detailed Microarchitecture Spec for the assigned block and maintain the Spec until its closure.
2	Partition the block into subblocks based on several factors, few being whether it is a Control, Datapath or Megacell, whether the block is too big to be Synthesized, whether breaking the block helps in resolving some of the floorplan and timing issues etc.
3	Come up with a Microarchitectural Spec for the Megacell. Present the Spec to the interested groups i.e mainly circuit designers who will be working on them.
4	Identify critical paths in the block and try to solve them through various design techniques i.e moving logic across stages, using complex gates etc. Talk to the circuit designer if anything fancy could be done (i.e using domino, dual rail circuits etc.) to solve the ones which cannot be solved using standard design techniques. If using such fancy techniques still results in timing violation then talk to the Architect to see what could be done next to solve the path.
5	Code RTL for the block and come up with a test bench environment to test it out. Convey to the full-chip/block-level Verification engineer the kind of tests you would be interested in.
6	Keep working towards a bug free design by constantly bombarding the design with various tests and debugging the failures.
7	Synthesize the control blocks and push them through various back-end flows.
8	Structurally code all the datapath blocks and push them through various back-end flows.
9	Run Static timing analysis on the block to make sure the block meets the timing spec.
10	Work with the Floorplan team in converging towards a workable floorplan for the block at the top level.
11	Keep track of the Area and Power number for the block and always thrive to be within the budget.
12	Present the Microarchitecture Spec to various groups within the team.
13	Patent all the novel design ideas/techniques which got implemented within the design which you were part off.

Other names associated with Logic Designer are RTL Designer or Microarchitect.

2. What is Max Timing and Min Timing?

Max Timing

Max time of a circuit is defined as the maximum amount of time data can take to traverse the logic between two memory elements (flops or latches). It determines the maximum frequency a circuit can operate at. Figure below shows a cycle of logic in a flop based design.

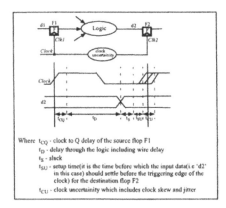

Figure 77: Cycle of Logic in a Flop Based Design

Equation for Max time with reference to figure above is given by,
    Max time $(t_{MAX}) = t_{CQ} + t_D + t_{SU} + t_{CU}$

Slack is given by, $t_S = t_{CP} - t_{MAX}$ (where t_{CP} is the clock period (i.e frequency of the Chip) and t_{MAX} is the Max time)

t_S can be a positive or negative number. A circuit with a negative slack is said to violate Max time (or otherwise known as Setup time violation).

Min Timing

Min time is the minimum amount of time data can take to traverse the logic between two memory elements (flops or latches). A Min time violation results in a circuit failure. Figure below shows a cycle of logic in a flop based design.

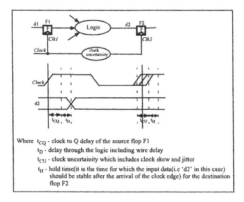

Figure 78: Cycle of Logic in a Flop Based Design

In the figure above, if $t_{CQ} + t_D < t_{CU} + t_H$ then we have a Min time violation.

An example of a Min time violation can be a signal going through minimum logic between two blocks in two separate clock skew domains (here t_{CU} would be high because of the two separate domains)

3. What is Power dissipation and what are some of the methods used to measure Power in a Chip?

Power dissipation is the amount of power dissipated in a circuit. There are three components to the amount of power dissipated in a circuit which are

1. *Static Power Dissipation* (P_S) - This is the power dissipated due to leakage current flowing through transistors which are off. It is given by the following equation,

$$P_S = \sum_{1}^{n} (\text{Leakage Current}) \times (\text{Supply Voltage})$$

Where 'n' is the number of devices

2. *Dynamic Power Dissipation* (P_D) - This is the power dissipated due to charging and discharging of load capacitance. This makes up a big chunk of the total power dissipated in a circuit. In most properly designed digital CMOS logic, the majority of the power goes into charging and discharging of load capacitance. These capacitors range in size from normal gate input and signal interconnect all the way up to main clock node on the Chip. These capacitors all get charged and discharged a certain number of times per clock cycle (often less than once per cycle) depending on the logic function and processor activity. The power dissipated here is given by the following equation,

$$P_D = CV^2F$$

Where, 'C' is the load capacitance
'V' is the supply voltage
'F' is the processor frequency

3. *Short Circuit Power Dissipation* (P_{SC}) - This is the power dissipated due to short circuit current (crowbar current). This is the current that flows from power to ground through a CMOS gate when its input voltage is not at VDD (Power) or VSS (Ground). This component of power is particularly sensitive to the slope of the gate's input signals. Slow rise/fall times at the input and fast fall/rise times at the output cause this component to be very high. It is given by the following equation,

$$P = VI$$

Where, 'V' is the supply voltage
'I' is the short circuit current

Total Power Dissipation which is the sum of the above three components is given by,

$$P_T = P_S + P_D + P_{SC}$$

Some important terms associated with Power are

Average Power Dissipation (P_{avg})	This is the power dissipated by a block when it is not idle and is processing a typical sequence of input vectors. *(This is typically measured over atleast 4 cycles; It effects the heat dissipated in the package during normal operation)*
Minimum Power Dissipation (P_{min})	This is the power dissipated by a block when it is idle (i.e block is inactive).
Maximum Power Dissipation (P_{max})	This is the power dissipated by a block in the cycle when the worst case sequence of input vectors are going through it. *(This is typically the worst case power in 1 cycle)*

Following are some of the methods used to measure Power in a Chip -

1. Prior to implementation Power Consumption of a Chip can be estimated using a formula based on the estimated number of gates (G), the capacitance of a typical gate (Cg), estimated number of wires (W), the capacitance of a typical wire (Cw), fre-

quency of the chip (F), the activity factor (A) and the voltage (V). The equation for Power Consumption here would be

$$P = (1.1GC_g + WC_w)\ V^2\ F\ A$$

In the above equation gate power is multiplied by 1.1 to account for the short circuit current (crowbar current) which in many cases is estimated to be 10% of the gate switching power. A typical gate in many cases is represented by a average sized 2-input NAND gate. Activity factor here is defined as the number of transitions per cycle divided by 2 *(activity factor for a static gate is assumed to be 0.25, for a dynamic gate is assumed to be 0.5 and for a clock it is 1 (since clock switches twice every cycle))*.

2. Prior to implementation Power Consumption of a Chip can be estimated based on area. The equation here typically used is,

$$P = (\text{Power density(W/mm}^2)\ \text{for Static Logic x Area occupied by Static Logic}) +$$
$$(\text{Power density(W/mm}^2)\ \text{for Dynamic Logic x Area occupied by Dynamic Logic}) +$$
$$(\text{Power density for(W/mm}^2)\ \text{Memory Structures x Area occupied by Memory Structures})$$

In the equation above most of the time the power density numbers are picked from previous processor projects and are scaled to the current process being used for the design.

3. Once the layout is ready, vectors may be used to measure power using one of the simulation tools (Simplex or Spice). The vectors chosen should be run for at least 4 cycles (this is a typical number used in the industry) to get a sustained average power. Normally the vectors are captured by running a power diag (developed by the verification folks with help from the designer) at the unit or chip level.

4. List some of the most common techniques used to improve timing of a Critical Path?

Table below lists some of the most common techniques used in the industry to improve timing of a critical path.

Table 24: Techniques to Improve Timing

1	Move logic across pipe stages. Figure below shows how moving logic across pipe stages helps in solving timing of a critical path. Here we have a fetch stage where instruction is accessed from I$ and a decode stage where instruction gets decoded to select either input 'a' or 'b'. As decoding and selecting either 'a' or 'b' results in a critical path we have moved portion of the decoding into fetch stage thereby solving the critical path. 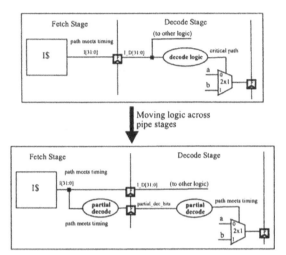 Figure 79: Moving Logic across Pipe Stages to Improve Timing
2	Move critical signal closer to the output of a Gate. In the figure below if we assume that input 'b' is critical then we would be better off in timing if we connect 'b' to the transistor closer to the output (i.e M3). The reason for this is when 'b' arrives the only charge it has to discharge is the charge at the output node against discharging charge at intermediate node N1 and output node when it is connected to M4. 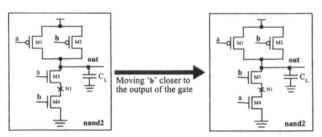 Figure 80: Moving Critical Signals Closer to the Output of a Gate to Improve Timing

Table 24: Techniques to Improve Timing

3	Move critical signal closer to the output. In the figure below if we assume 'a' to be critical when compared to the rest of the inputs then we will be better off in timing if we move 'a' closer to the output.

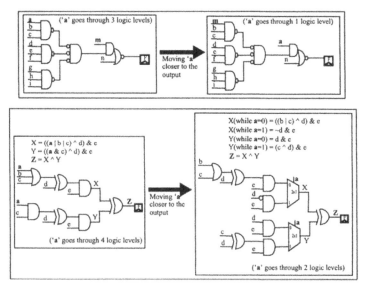

Figure 81: Moving Critical Signal Closer to the Output to Improve Timing

4	Replicate portion of the logic in the path which is timing critical. Figure below shows how replicating logic resolves our critical path.

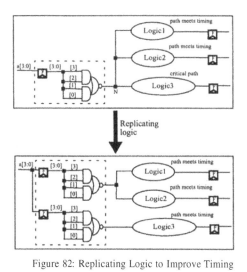

Figure 82: Replicating Logic to Improve Timing

Table 24: Techniques to Improve Timing

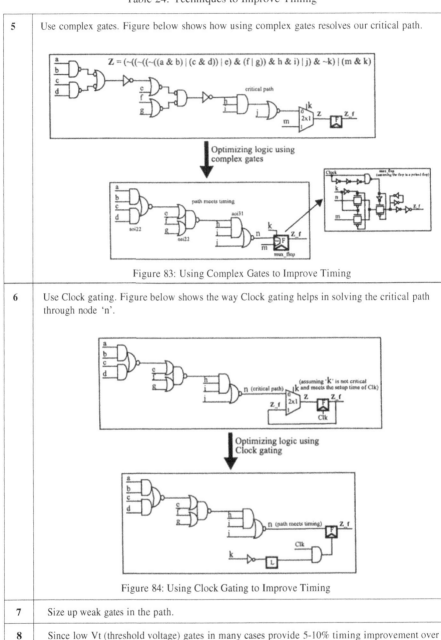

5	Use complex gates. Figure below shows how using complex gates resolves our critical path.

Figure 83: Using Complex Gates to Improve Timing

6	Use Clock gating. Figure below shows the way Clock gating helps in solving the critical path through node 'n'.

Figure 84: Using Clock Gating to Improve Timing

7	Size up weak gates in the path.
8	Since low Vt (threshold voltage) gates in many cases provide 5-10% timing improvement over standard Vt gates, use low Vt gates in the paths which are timing critical.

Table 24: Techniques to Improve Timing

9	If the path is wire dominated then do the following things -

a. Increase the width of the wire as this would reduce the wire resistance thereby improving timing.

b. Increase spacing between the wire under reference and its neighbors as this would reduce coupling capacitance thereby improving timing.

c. Single/Double shielding the wire under reference would reduce coupling capacitance thereby improving timing.

d. Balancing wire load by inserting a repeater helps in getting a better edge rate at the destination (i.e node A in the figure below) thereby improving timing. Figure below shows how inserting a repeater helps in resolving the critical path.

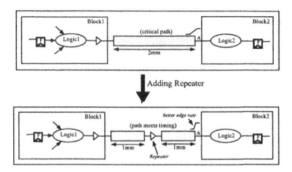

Figure 85: Adding Repeater to Improve Timing

e. Use a higher metal layer (i.e if the current route is in Metal2 then try using higher Metal layer which is less resistive when compared to Metal2) for the route as this would result in reduced wire resistance thereby improving timing.

f. Tapering wire improves timing as doing this reduces the total wire capacitance.

Figure 86: Tapering Wire to Improve Timing

10	Use non-static logic (i.e domino, dual rail, pseudo nMOS etc.) instead of static logic. This should be the last resort because of the following disadvantages

a. Non-static circuits tend to dissipate more power when compared to static.

b. Doesn't scale well when moving to a better process.

c. Needs more design time when compared to static.

Table 24: Techniques to Improve Timing

11	Using a pulse flop instead of a master-slave flop. Since pulse flop tends to have a negative setup time when compared to a master-slave flop which has a positive setup time (typically 40-60ps without any embedded gates in 0.09micron process technology), replacing master-slave with a pulse flop at the destination helps in solving some of the timing paths. Pulse flop typically has higher clock to Q and power dissipation when compared to a master-slave flop, so we need to be careful here to make sure the logic in Stage2 (in figure below) also meets timing with pulse flop in place. Figure below shows how replacing a master-slave flop with a pulse flop resolves our critical path. 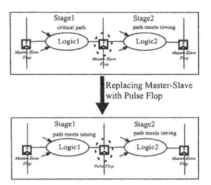 Figure 87: Replacing Master-Slave Flop with a Pulse Flop to Improve Timing
12	At an architectural level timing of instruction decode logic could be improved by generating predecode bits for an instruction on its way from L2/Memory and storing it in I$ along with the instruction. Figure below shows how generating predecode bits and storing it in I$ resolves our critical path in Decode stage. 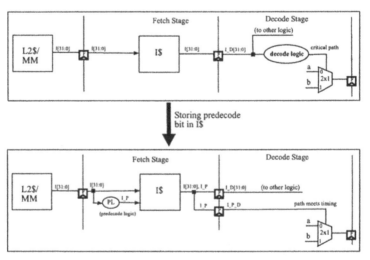 Figure 88: Storing Predecode Bits to Improve Timing

5. List some of the most common techniques used to fix Mintime Paths in a Design?

Table below lists some of the most common techniques used in the industry to fix Mintime Paths in a Design.

Table 25: Techniques to Fix Mintime Paths

1	Use Poly or High Resistive Metal in the Path. Use this approach if the block is Gate dominated. Figure below shows how using a poly or high resistive metal helps in fixing the Mintime path. 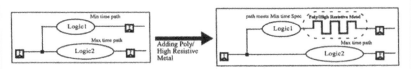 Figure 89: Using a Poly or High Resistive Metal to Fix Mintime Path
2	Use a Mintime flop (Flop with higher Clock to Output delay (i.e C to Q delay)) in the Path. Figure below shows how using a Mintime flop helps in fixing the Mintime path. 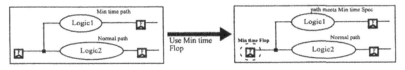 Figure 90: Using a Mintime Flop to Fix Mintime Path
3	Add Mintime buffer (i.e buffer designed to take care of the Mintime violation (i.e delay through the buffer ensures that for the case of a back to back flop with minimum distance the placement of this buffer in its path avoids any mintime violation)) in the Path. We need to be careful here as this might result in an increase in area. Figure below shows how adding a Mintime buffer helps in fixing the Mintime path. 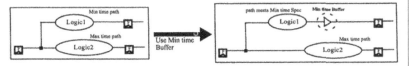 Figure 91: Adding a Mintime Buffer to Fix Mintime Path
4	Down size logic in the Path. We need to be careful to make sure that the slew rate of intermediate nodes within the Path doesn't result in any noise violations. Figure below shows how downsizing logic helps in fixing the Mintime path.

Table 25: Techniques to Fix Mintime Paths

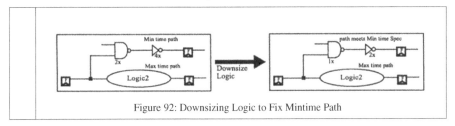

Figure 92: Downsizing Logic to Fix Mintime Path

6. List some of the basic techniques used to reduce Power Dissipation at the Logic/Circuit level?

Table below lists some of the most common techniques used in the industry to reduce Power dissipation at the Logic/Circuit level.

Table 26: Techniques to Reduce Power Dissipation

1	Provide support for Clock gating. Clock gating is a technique of masking off the clock when circuits are idle and thus significantly reducing the switching activity in a circuit and on the clock nets. Since clocks contribute to 30-40% of the total power dissipation, Clock gating is considered as the most effective means of reducing power dissipation. In figure below by AND-ing the clock with Clock_Enable control signal we essentially disable the clock from driving the Flop whenever needed thereby saving power.

Figure 93: Clock Gating

Figure below shows Clock gating as applied to a 1-Way Superscalar Pipelined Processor. Here we turn off the clock for few flops in E stage if we find that there is no ALU related instruction in R stage. Here we need to latch the enable signal before gating it with Clk in order to avoid any glitches in GClk.

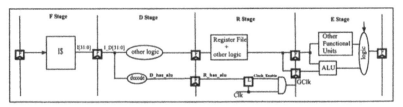

Figure 94: Clock Gating as Applied to a 1-Way Superscalar Pipelined Processor

Table 26: Techniques to Reduce Power Dissipation

| 2 | Forwarding an extra bit with the Data or Control bus which indicates whether the bus has valid data or not. The flops at the receiving end of the bus only flop the data if the extra bit is set. Doing this will allow logic beyond the receiving flops (i.e logic in Stage2 in the figure below) to be inactive while the extra bit is not set thereby saving power. This method of saving power in some cases is known as Logic gating. In figure below P1 < P0. |

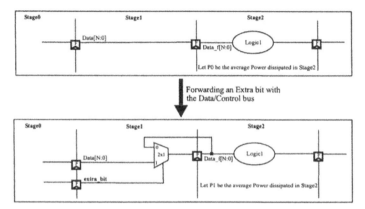

Figure 95: Logic Gating

Figure below shows Logic gating as applied to a 1-Way Superscalar Pipelined Processor. Here we prevent signal toggling at the output of few flops in E stage by recirculating the data while there is no ALU related instruction in R stage.

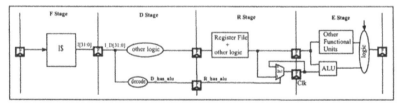

Figure 96: Logic Gating as Applied to a 1-Way Superscalar Pipelined Processor

Table 26: Techniques to Reduce Power Dissipation

3	Provide support for Power gating. As we go into deep submicron technology power dissipated due to leakage current is of a great concern as current due to leakage increases in deep submicron technology because of the lowered threshold voltage. One way to reduce leakage is by providing support for Power gating. The idea is to introduce an extra transistor in the supply voltage or ground path. The extra transistor is turned on in the used section and turned off in the unused section. Figure below shows a power gated NAND circuit. In the circuit below Power_Enable control signal is set to logic0 when the NAND gate is used/active and set to logic 1 when the NAND gate is unused/idle.

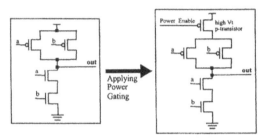

Figure 97: Power Gating

Figure below shows Power gating as applied to a BIST Controller. Since BIST controller is active only while we are in BIST mode, here we cut off the power to the entire controller while in non-BIST mode by applying inverted version of the 'bist_mode' signal as input to the additional P-transistor.

Figure 98: Power Gating as Applied to a BIST Controller

4	Determine signals with high activity levels and try to optimize paths involving these signals so that these high active signals go through minimum logic levels. Doing this will reduce power in these paths involving high active signals. Figure below shows the way path involving highly active signal 'a' is optimized.

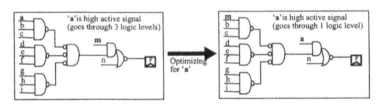

Figure 99: Optimizing Paths with Highly Active Signals to Reduce Power

5	Use high Vt (threshold voltage) gates in non-critical paths as they dissipate less power when compared to standard Vt gates.

Table 26: Techniques to Reduce Power Dissipation

| 6 | Since bigger gates consume more power, reducing the size of gates in non critical paths results in reduced power dissipation. Figure below shows the way gates in a non-critical path could be downsized thereby reducing power dissipation in the path but still meeting timing. In the figure below ?x represents the drive strength of a gate where gate with a drive strength of 1x is weaker than a similar gate with a drive strength of 2x and so on and so forth. |

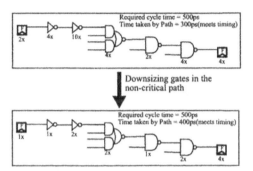

Figure 100: Downsizing Gates to Reduce Power

| 7 | Try to achieve better edge rates on signals as slow edge rates results in DC power consumption in the receiver. This is because if a signal takes too long to switch then the receiver which is sensing the signal draws excessive current as the pullup and pulldown transistors of the receiving device are partially on all the time during which the switching of the signal is happening. In the figure below since t_{rs2} (transition time of signal A at Inverter N) < t_{rs1} (transition time of signal A at Inverter M), M1 and M2 of Inverter N are partially ON for a less amount of time when compared to that of Inverter M as a result of which DC power dissipation (because of short circuit current) in N is less than that in M, so P1 < P0. (Here we have assumed that increasing the size of the inverter at the source in order to get a better edge rate has minimal effect on the Power number in the path) |

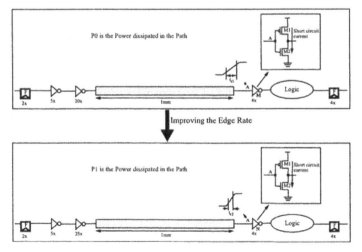

Figure 101: Improving Edge Rate to Reduce Power

Table 26: Techniques to Reduce Power Dissipation

8	Since Clock accounts for 30-40% of the total power dissipation, reducing loading on the clock network will result in reduced power dissipation. One example would be using register file structures instead of flops where ever possible.
9	Making sure that the wiring between gates is kept to minimum. Doing this will reduce the wire capacitance and thereby results in reduced power dissipation.
10	Since non-static circuits consume more power avoid using non-static circuits (i.e domino, dual-rail, pseudo-nMOS etc.) in the design.
11	Encode the Bus before transmitting it long distance. A data bus such as a counter value (in the figure below sq_id[3:0] (4-bit store queue ID) which is assumed to always increment by 1/0 in any given Cycle) which always steps by 1 will greatly benefit in terms of Power dissipation if the value is encoded to Gray code before transmission and then decoded back to Binary format at the receiving end. In the figure below P1 < P0. The benefit in Power dissipation by encoding the value to Gray code comes from the fact that only one of the four bits will be changing on the 4-bit bus (sq_id[3:0] in the example below) as we go from Cycle to Cycle as against more than one bit changing if the data were to be forwarded without encoding. 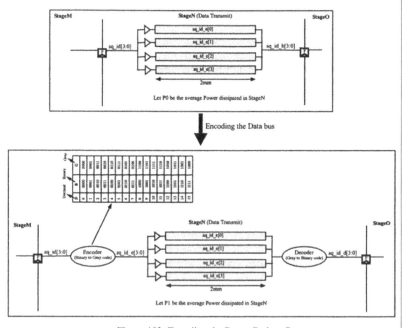 Figure 102: Encoding the Bus to Reduce Power
12	Allow some of the non critical flops to be non scannable as this would result in less logic within the flop thereby saving power.

7. What is a Synchronizer and what are some of the design guidelines used to help data integrity in designs with multiple clock domains?

Synchronizer circuits are needed in a multi clocked domain system where a signal generated in one clock domain feeds into a memory element (typically Flop) in a different clock domain. Synchronizers help in preventing metastability at the destination flop. Metastability is a state of a memory element where the output of the memory element stays near VDD/2. Few ways of preventing metastability are

1	Insert a synchronizer in a signal path before entering the new clock domain. This method is typically used for control signals.
	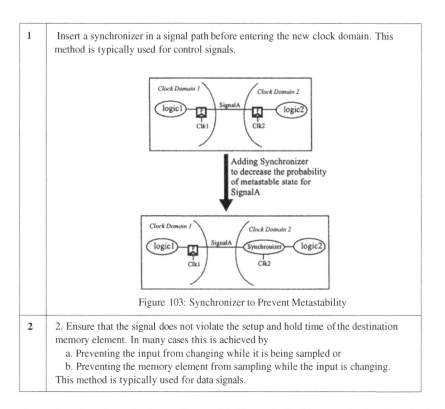
	Figure 103: Synchronizer to Prevent Metastability

2	2. Ensure that the signal does not violate the setup and hold time of the destination memory element. In many cases this is achieved by
	    a. Preventing the input from changing while it is being sampled or
	    b. Preventing the memory element from sampling while the input is changing.
	This method is typically used for data signals.

A synchronizer is typically made up of 2 flops clocked by the destination clock. Unlike regular flops, these flops are specially designed to exit the metastable state quickly. In many cases these flops are known as metastability hardened flops. Figure below shows a 2-flop Synchronizer.

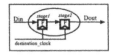

Figure 104: 2-flop Synchronizer

In the figure above the output of the first flop in a synchronizer could be in a metastable state (as the input to synchronizer could arbitrarily change) for a period of time but by the time the signal reaches the second flop there is a high probability that the metastable state would have been resolved to a logic 0 or logic 1. There is still a finite probability that the metastability would not have resolved in which case we have a synchronization failure. The goal in a synchronizer design is to minimize the mean time between failures (MTBF) to an acceptable level. MTBF is given by the following equation

$$MTBF = \frac{e^{t_r/t_{switch}}}{f_c f_d T}$$

Where t_r - time which the output of the latch must be settled to a
logic0 or logic1

t_{switch} - time constant for transition from a metastable state to a valid state

T - time the data input requires to pass through the V_{IH} and V_{IL} regions

$$T = \frac{(V_{IH} - V_{IL})}{\frac{d(Vin)}{dt}}$$

input high noise margin   input low noise margin

rate of change of data input

f_c - Clock frequency

f_d - Data frequency

In the above equation

a. Allowing more resolution time (i.e 't_r') is probably the simplest way to exponentially reduce the failure rate (see if the output of the synchronizer can be avoided from being sampled on every clock cycle; do not have any logic between the 2 flip-flops in a Synchronizer).

b. Decreasing 'f_c' and 'f_d' helps in reducing the failure rate (in cases where the data does not have to be sampled on every clock cycle, synchronizer could be clocked at half the frequency of the receiver clock).

c. Decreasing 'T' helps in reducing the failure rate (careful design of the latch input stage helps minimize T. Providing clean, crisp inputs can also help reduce this danger window.

d. Decreasing 't_{switch}' of the regenerative feedback loop will help in exponential decline in failure rate. Keeping the parasitic load on the critical nodes as low as possible and using small devices in the feedback loop and buffering the flop output helps in decreasing 't_{switch}' factor.

Some of the design guidelines used to help data integrity in designs with multiple clock domains are

1. Partition the design carefully to minimize the number of synchronizers used.

2. Input to a synchronizer should come from a flop or guaranteed to be monotonic (if the input to a synchronizer comes via combinational logic then it is likely that the signal may transition multiple times during a cycle before settling on a final value (due to variations in the timing paths through the combinational logic) which may cause false information to pass through the synchronizer which can be fatal).

3. In the case of non-stream data inputs crossing a clock domain (i.e going from one clock domain to a different clock domain), cover the inputs with a strobe i.e place the data on the data bus and assert a strobe. Here the strobe is fed through a synchronizer to the destination clock domain whereas the data on the data bus is left until it is guaranteed that the data has been consumed by one of the following methods

a. The destination sends back an acknowledge signal through a synchronizer indicating that the data has been consumed (this process is known as flow control mechanism).

b. If you know the clock ratios, maximum latency through the synchronizer and the maximum latency between the strobe signal arrival and the last use of data, then use this information to appropriately control your data flow. Figure below shows a typical usage of synchronizer for non-stream data.

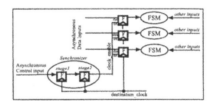

Figure 105: Synchronizing Non-Stream Data through a Strobe

4. In the case of a stream data crossing clock domains, instead of synchronizing each and every unit of data from one clock domain to another, use a FIFO (First In First Out) memory structure which uses clocks from both source and destination clock domains. This minimizes the number of synchronization events needed resulting in better reliability and performance. Here logic in the reading clock domain only needs to know when there is data in the FIFO to be read and when it is empty whereas logic in the writing clock domain only needs to know if the FIFO is full or not. Here the FIFO full and empty conditions are correctly managed by forwarding the write pointer to the destination clock domain and the read pointer to the source clock domain through synchronizers after going through encoding. Gray coding is the most commonly used encoding where if a pointer has advanced by two, the only possible outputs from the synchronizer would be advance by zero, one or two in which case any of these outputs will allow correct operation. Likewise if the pointer has advanced by three, the only possible outputs from the synchronizer would be advance by zero, one, two or three thereby allowing correct operation. Figure below shows this scheme.

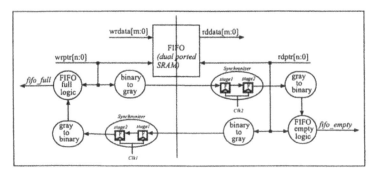

Figure 106: Synchronizing Stream Data through FIFO Structure

5. When sending data from high clock domain to low clock domain either provide high bandwidth (enough synchronizers must be banked to reduce each synchronizer's input pulse transitions to a low enough rate to allow each synchronizer to work correctly at the given clock ratio) or slow down the data (by providing flow control mechanism). Figure below shows a typical synchronization logic while transferring data from high clock domain to lower clock domain.

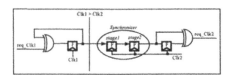

Figure 107: Synchronization Logic for Data Transfer from Higher Frequency Clock Domain to Lower Frequency Clock Domain

8. What are some of the characteristics of a Flop and a Latch?

Table below shows the characteristics of a Flop and a Latch.

Table 27: Characteristics of a Flop and a Latch

Flop	Latch
1. It is edge triggered. Figure below shows a Pulse Flop. Figure 108: Pulse Flop	1. It is level sensitive. Figure below shows a Level Sensitive Latch. Figure 109: Positive Level Sensitive Latch

Table 27: Characteristics of a Flop and a Latch

Flop	Latch
2. There is no time borrowing. Here the data must arrive at least a setup time before the active edge of the Clock. This limits the amount of logic in a given cycle. Figure below shows the maximum logic delay in a flop based design. Figure 110: Maximum Logic Delay in a Flop Based Design	2. We can borrow time from the next phase or the previous phase. Here time borrowing occurs when data changes while the latch is transparent. Time borrowing allows a circuit to use time from a previous or next phase to perform its logic. Figure below shows forward time borrowing in a latch based design. Figure 111: Time Borrowing in a Latch Based Design
3. Involves more logic resulting in more area and power.	3. Involves less logic resulting in less area and power.
4. Flop based designs are more tool friendly than latch based designs.	4. Latch based designs are less tool friendly than flop based designs.

9. Describe with an example a Picker Logic associated with an Issue Queue in an Out-Of-Order Processor?

Picker logic picks instructions to be issued for execution. Most of the time these instructions are sitting in a buffer structure before they get picked. Issue Queue in an Out-Of-Order Processor uses picker logic to pick instructions for execution by one of the functional units. *(other places where picker logic gets used are, Load Queue uses picker logic to pick a load among bunch of loads sitting in the Load Queue for cache access, Store Queue uses picker logic to pick a store among bunch of stores sitting in the Store Queue for cache access etc.)*

Figure below shows Picker Logic for an 8-entry Issue Queue (IQ) in an Out-Of-Order Processor. IQ is the place holder for instructions before they get picked for execution. In the figure below, WV[7:0] is the Wrap Vector (maintained by the IQ *(wrap bit could also be forwarded by one of the blocks preceeding IQ involved in assigning instruction ID's)*) where in case of an instruction wrap around in the IQ, the wrap bit corresponding to that instruction is toggled as can be seen from the figure below, RV[7:0] is the Ready Vector (maintained by the IQ) where a bit in the vector gets set when its corresponding instruction is ready to be issued (i.e it is dependent free). Based on the Wrap Vector and Ready Vector one of the instructions sitting in the IQ gets picked for issue (i.e to the execution unit). Here we see that if multiple instructions are ready then the oldest gets picked for issue.

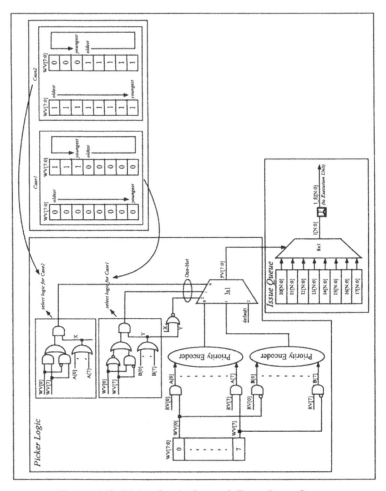

Figure 112: Picker Logic for an 8-Entry Issue Queue

10. What are some of the advantages and disadvantages of implementing Mux functionality using an AOI against a Pass Gate?

Table below lists some of the advantages and disadvantages of implementing Mux functionality using an AOI against a Pass Gate.

Table 28: AOI against Pass Gate

Mux functionality using AOI	Mux functionality using Pass Gate
1. We don't have to worry about signal contention while in scan mode. 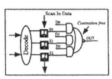 Figure 113: Figure 1 for AOI	1. Need to make sure Mux exclusivity is guaranteed in scan mode otherwise would result in signal contention thereby damaging the device. 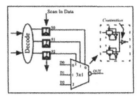 Figure 114: Figure 1 for Pass Gate
2. In normal functional mode we don't have to worry about the arrival times of select controls (i.e S0, S1 and S2) to an AOI. 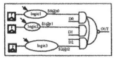 Figure 115: Figure 2 for AOI	2. In normal functional mode we do have to worry about the arrival times of select controls to a Mux. The reason being if the final value of {S2S1S0} in figure below is 3'b001 but S0 changes to 1 at time t0 and S2 changes to 0 at time t2 and (t2-t0) is a big number then contention on the internal node will result in damaging the device over a period of time. So here we need to make sure that all selects to the Mux do come within a fixed interval of time. 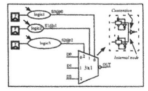 Figure 116: Figure 2 for Pass Gate

Table 28: AOI against Pass Gate

Mux functionality using AOI	Mux functionality using Pass Gate
3. Provides better ATPG coverage than Pass Gate implementation in certain implementation scenarios. In figure below where the decode logic for the selects is on the other side of the flops, we still get 100% ATPG coverage through the AOI gates implementing a Mux functionality.	3. Less ATPG coverage than AOI implementation in certain implementation scenarios. In figure below because of the Mux protection logic (to take care of contention while in scan mode) in the select path we won't get 100% ATPG coverage through the Mux as the paths through D0, D1 and D2 are never selected in scan mode.

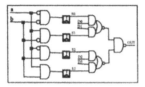

Figure 117: Figure 3 for AOI

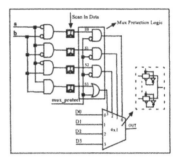

Figure 118: Figure 3 for Pass Gate

| 4. Smaller width Mux functionality implemented using AOI's tend to give better timing number than implementing the same functionality using a Pass Gate. | 4. Wide input Mux functionality implemented using a Pass Gate tend to give better timing number than implementing the same functionality using AOI's. |

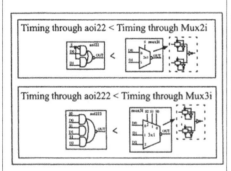

Figure 119: Figure 4 for AOI

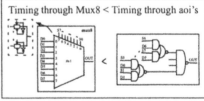

Figure 120: Figure 4 for Pass Gate

Table 28: AOI against Pass Gate

Mux functionality using AOI	Mux functionality using Pass Gate
5. In figure below where the logic uses an AOI ('aoi22'), the timing through path P2 depends only on the arrival time of 'D1'. 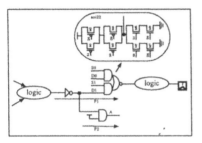 Figure 121: Figure 5 for AOI	5. In figure below where the logic uses a Pass Gate Mux ('mux2i'), the timing through path P2 depends on the following events - a. Arrival time of 'D1', if 'D1' arrives after the arrival of 'S0' and 'S1' (i.e by the time 'D1' arrives 'S0' and 'S1' are in a valid mutually exclusive state). b. Arrival time of 'S0' and 'S1', if 'S0' or 'S1' arrives after the arrival of 'D1'. If 'S0' or 'S1' arrives after the arrival of 'D1' and there is a period of time during which they are non mutually exclusive after the arrival of 'D1' then during that period node 'A' will be in some non-deterministic state. The value will settle down to logic0 or logic 1 (based on the value of 'D1') once the selects (i.e 'S0' and 'S1' here) exit the non mutually exclusive state. So here even though 'D1' arrives early the delay through path P2 is really determined by the arrival times of 'S0' and 'S1'. As an example if 'D1' arrives at time 't0', 'S1' arrives at 't1' and 'S0' arrives at 't2' and t2>t1>t0 then node 'A' will be in some non-deterministic state if 'D0' and 'D1' have opposite values while 'S0' and 'S1' are in non mutually exclusive state and will go to a stable state after the arrival of 'S0'. So here we see that the path through node 'A' is really dependent on the arrival time of 'S0' 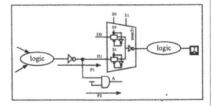 Figure 122: Figure 5 for Pass Gate

11. Describe with an example a Working Register File ID Assignment Logic for an Out-Of-Order Processor supporting Multiple Threads?

In an out-of-order processor, there exists a Working Register File (WRF)/Re-Order Buffer where the results of uncommitted instructions (i.e instructions which are not yet committed) go and sit before being written into the architectural register file at the time of commit. Any valid instruction (typically with a valid destination register) in an out-of-order processor gets renamed to one of the entries in WRF allowing the instruction to be issued out-of-order. This renamed value assigned to an instruction is known as Working Register File ID (WRF_ID). Logic involved in assigning WRF_ID's is known as WRF_ID assignment logic. Working register file allows an out-of-order processor to take advantage of instruction level parallelism in the code whereby the processor can issue an younger instruction which is dependent free before an older instruction gets issued.

A multithreaded processor (i.e a processor supporting multiple threads) could be supporting any one of the following thread switching algorithms which are VT (Vertical Threading), SMT (Simultaneous Multithreading), HT (Horizontal Threading), PT (Power Threading) and BT (Branch Threading). Here lets assume the processor to be supporting VT. Also lets assume the out-of-order processor to be a N-stage pipelined, 3-Way superscalar processor supporting two threads/strands (thread and strand are synonyms and have been used interchangeably in the following paragraphs). Figure 30 below shows three of its N stages.

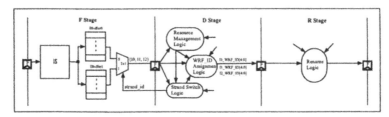

Figure 123: Fetch, Decode and Rename Stage of a N-Stage Pipelined Processor

The three stages shown are Fetch (F), Decode (D) and Rename (R). Since the processor supports dual strands we have two instruction buffers, one for each strand as shown in Figure 123. In F stage we fetch instructions for each strand and feed them into the appropriate instruction buffers. Based on the stand ID information from D stage, instructions from appropriate strand gets forwarded to D stage. Lets assume the WRF_ID Assignment Logic along with the main Strand Switching Logic (SSL) and Resource Management Logic (RML) to be part of the D stage. SSL provides the appropriate strand related information to WRF_ID assignment logic. It takes inputs from Resource Management Logic (which manages various resources in various units down the pipe), Commit Unit (which commits instructions based on the completion and exception report) and Fetch Unit (which fetches instructions in addition to managing the instruction buffers for individual strands). RML provides information about the stall condition (i.e stall while there is a resource scarcity (i.e one of the resources down the pipe is full or close to its high threshold value)) for appropriate strand, Commit Unit provides the flush information (i.e in the case of a branch mispredict in a particular strand or a particular strand being parked or an instruction in a particular strand resulting in an exception etc.) for appropriate strand and Fetch Unit provides the instruction buffer empty information for each strand.

Figure 124 below shows the pipeline diagram showing fetch groups from Strand0 and Strand1 while there is a strand switch and resource stall. We see from the figure that while there is a strand switch there is no penalty (i.e there is no bubble in the pipe (i.e in the following cycle you see instructions from the other Strand in D stage)).

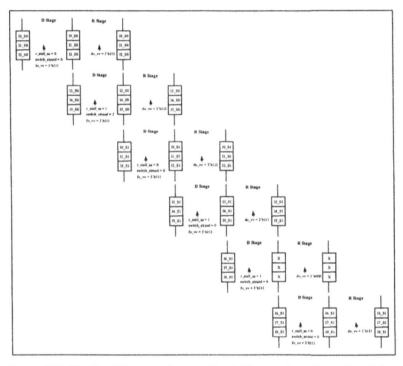

Figure 124: Pipeline Diagram showing Fetch Groups from Strand0 and Strand1

Definition of various terms used in Figure 124 above are

r_stall_as - resource stall for the strand actively being processed by D stage.

switch_strand - it refers to the condition where you switch strands i.e 'switch_strand=1' means there will be instructions from the other strand(i.e strand other than the current active strand) in D stage next cycle and 'switch_strand=0' means D stage will see instructions from the current active strand next cycle.

fu_vv[2:0] - valid vector corresponding to the instructions in D stage. It is the floped version of valid vector forwarded by F stage.

du_vv[2:0] - valid vector corresponding to the instructions in R stage. It is the floped version of valid vector forwarded by D stage.

I*_S0 - instruction belonging to Strand 0

I*_S1 - instruction belonging to Strand 1

Let's assume the processor to support a 32-entry WRF structure, a 32-entry Issue Queue and a 32-entry Commit Queue. Also let's assume the WRF, Issue Queue and Commit Queue to be a 32 entry structure in single strand mode and is split between strands (i.e 16 entries per strand) in dual strand mode. Figure 125 shows a flow process for assigning WRF_ID's for such a processor. Figure 126 shows the logic involved in assigning WRF_ID's for such a processor. Here the working register file ID's (WRF_ID's) for individual instructions are assigned by maintaining an active strand pointer (AS_PNTR[4:0]) and a set of individual strand pointers (S0_PNTR[4:0], S1_PNTR[4:0]). The active strand pointer is maintained based on data pertaining to instructions from a strand actively being processed by the Decode Unit, while the individual strand pointers are maintained based on data pertaining to instructions from a relevant strand. When a fetch group (i.e a maximum of 3 valid instructions (since it is 3-Way Superscalar)) enters D stage, both the active strand pointer and the individual strand pointer for the relevant strand are updated based on its valid vector and other information as shown in Figure 126. Logic in Figure 126 is based on the pipeline diagram and flowchart shown in Figures 124 and 125. WRF_ID assignment logic takes necessary strand related information from SSL. Necessary strand related information being, current strand being processed, whether we will be switching strands in next cycle etc. Here every instruction gets assigned a WRF_ID irrespective of whether it has a valid destination register or not. Since Commit Queue is 32 entries deep assigning WRF_ID's to every valid instruction doesn't hurt our performance. If we were to assign WRF_ID to valid instructions with valid destination registers only then it could be achieved by making few updates to the logic shown in Figure 126. Definition of various terms used in Figure 126 are

AS_PNTR[4:0] - 5-bit Active Strand pointer.

S0_PNTR[4:0] - 5-bit Strand0 pointer.

S1_PNTR[4:0] - 5-bit Strand1 pointer.

flush_pipe_s0 - flush data corresponding to Strand0. This signal comes from Commit Unit.

flush_pipe_s1 - flush data corresponding to Strand1. This signal comes from Commit Unit.

flush_pipe_as - flush data corresponding to current active strand in D stage. This is generated in D stage based on strand information from SSL and flush pipe information for individual strands from Commit Unit.

RESET - global reset.

current_strand - strand actively being processed in D stage. 'current_strand=0' means the strand actively being processed in D stage is Strand0 and 'current_strand=1' means the strand actively being processed in D stage is Strand1.

next_strand - strand that will be processed in D stage in the next cycle. 'next_strand=0' means the strand that gets processed in the next cycle in D stage is Strand0 and 'next-strand=1' means the strand that gets processed in the next cycle in D stage is Strand1.

single_strand_mode - while active indicates that only one strand is currently active in the CPU core. Here the active strand could be either Strand0 or Strand1.

switch_strand - It refers to the condition where you switch strands i.e 'switch_strand=1' means there will be instructions from the other strand(i.e strand other than the current active strand) in D stage next cycle and 'switch_strand=0' means D stage will see instructions from the current active strand next cycle.

fu_vv[2:0] - Valid vector corresponding to the instructions in D stage. It is the floped version of valid vector forwarded by F stage.

r_stall_as - Resource stall for the strand actively being processed by D stage.

I*_WRF_ID[4:0] - renamed instruction ID's for the individual instructions.

Logic A in Figure 126 is used to take care of the condition where you are not switching strands but there is a resource stall on the current active strand as a result of which you should not be incrementing your counters next cycle (assuming resource stall for the active strand is still active) as no valid instruction bundle will be forwarded next cycle. Logic B in Figure 126 is used to take care of the condition where you switched to Strand0 but Strand0 is already in stall mode or you switched to Strand1 but Strand1 is already in stall mode in which case you should not be incrementing your counters as there will be no valid instruction bundle forwarded in the current cycle.

One thing to note here is that in dual strand mode we forward 'cs' (current active strand bit which is '0' if the strand actively being processed in D stage is Strand0 and '1' otherwise) in the least significant bit position of the WRF_ID. The reason for doing this is, here if we assume that Issue Queue in I stage (not shown in Figure 123) uses WRF_ID to update its entries with the new incoming instructions, having 'cs' bit in the least significant bit position of the WRF_ID in dual strand mode helps Issue Queue maintain fairness between strands to be picked for issue in certain implementations. One such implementation would be where Issue Queue is a unified structure and the priority encoder used to pick an instruction picks in a circular fashion based on age and ready status. As an example lets assume that there are 3 priority encoders (i.e PE0 for Slot0, PE1 for Slot1 and PE2 for Slot2) to pick three instructions stationed in Issue Queue for each of the 3 execution slots (since processor is 3-way superscalar) and also lets assume that IDU is currently processing instructions from Strand0. If we assume that IDU assigns instructions in the current fetch group (i.e I0, I1 and I2) from Strand0 to Slot0 with WRF_ID's 0,2 and 4, then these instructions go and sit in entries 0, 2 and 4 in the Issue Queue. Now if IDU switches strands (i.e switches to Strand1) and processes fetch group from Strand1 then lets assume that IDU assignees WRF_ID's 1,3 and 5 to instructions I0, I1 and I2 from Strand1. These instructions from Strand1 go and sit in entries 1, 3 and 5 in the Issue Queue. If in a given cycle say N, I0 (belonging to Strand0) in entry 0, I0 (belonging to Strand1) in entry 1 and I1 (belonging to Strand0) in entry 2 are ready to be picked for issue, then PE0 picks I0 (belonging to Strand0) to be issued in cycle N, picks I0 (belonging to Strand1) to be issued in cycle (N+1) and picks I1 (belonging to Strand0) to be issued in cycle (N+2). So here we see that we have provided fairness in picking instructions belonging to different strands by forcing instructions for Strand0 to go and sit in even entries of Issue Queue and instructions for Strand1 to go and sit in odd entries of Issue Queue by having 'cs' bit as the least significant bit of the WRF_ID in dual strand mode.

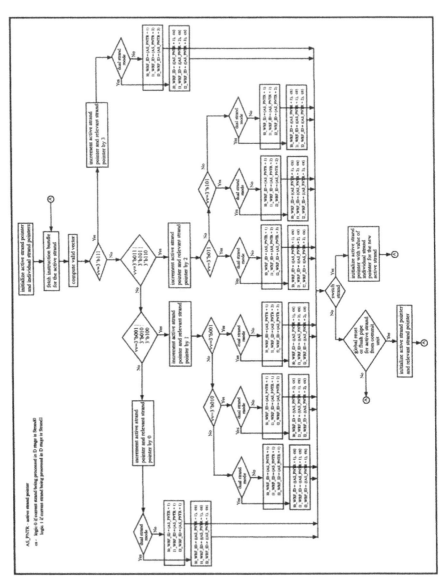

Figure 125: Flow Process for Assigning WRF_ID's in an Out-Of-Order Processor
Supporting Dual Strands

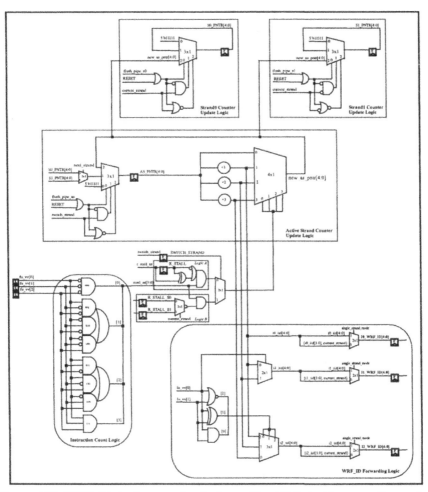

Figure 126: WRF_ID Assignment Logic for an Out-Of-Order Processor Supporting
        Dual Strands

This page intentionally left blank

3 Circuits and Layout

1. What are some of the responsibilities of a Circuit Designer in the Chip Industry?

Some of the responsibilities of a Circuit Designer are summarized below.

Table 29: Circuit Designer Responsibilities

1	Design some of the key circuit intensive elements on the Chip such as SRAM's, Register Files, ROM's, PLA's, Buffer/Queue structures, Library Cells, PLL, DLL, CAM's etc.
2	Work on feasibility studies of Critical Paths.
3	Characterize various Megacells and Library cells.
4	Come up with a detailed Spec for the Megacell, work on designing the Megacell and run dynamic simulation and formal verification to check equivalence between the Megacell and its behavioral model.
5	Run Min time and Max time checks (i.e Hold time and Setup time violation checks) on the final Megacell layout, analyze the reports and fix any violations.
6	Run Noise check (check for Noise violations) on the Megacell/Library cell layout, analyze the report and fix any violations.
7	Run EM/IR check (Electromigration and Voltage drop check)) on the Megacell/Library cell layout, analyze the report and fix any violations.
8	Run Clock check (check for Clock network meeting the Clock Spec) on the Megacell/Library cell layout, analyze the report and fix any violations.
9	Patent any novel circuit techniques which you were part off and which got implemented in the Chip.

2. What are some of the responsibilities of a Physical/Integration Designer in the Chip Industry?

Some of the responsibilities of a Physical/Integration Designer are summarized below.

Table 30: Physical/Integration Designer Responsibilities

1	Custom Layout of Transistors and Gates.

Table 30: Physical/Integration Designer Responsibilities

2	Custom Routing of signals between Transistors, between Gates and between Blocks within a Chip
3	Running LVS (Layout Versus Schematic), DRC (Design Rule Checking) and ERC (Electrical Rule Checking) on the Layout. Analyzing the reports to see if there any errors and if there are then fixing them in the Layout.
4	Custom Routing of Power grid and Clock grid.
5	Work on Unit/Chip Floorplan.
6	Run Min time and Max time checks (i.e Hold time and Setup time violation checks) at the Unit and Chip level, analyze the reports and fix any violations.
7	Run Noise check (check for Noise violations) at the Unit and Chip level, analyze the report and fix any violations.
8	Run EM/IR check (Electromigration and Voltage drop check) at the Unit and Chip level, analyze the report and fix any violations.
9	Run Clock check (check for Clock network meeting the Clock Spec) at the Unit and Chip level, analyze the report and fix any violations.
10	Patent any novel Layout techniques which you came up with and got implemented in the Chip.

3. What is Resistance, Capacitance and Inductance?

Table below provides definitions for Resistance, Capacitance and Inductance.

Table 31: Resistance, Capacitance and Inductance

Resistance	All materials impede the flow of current to some extent. This property is called Resistance. Insulators have a very high resistance to current flow whereas Conductors have low resistance and allow current to flow freely. The unit of measure for Resistance is Ohms. Resistance of a wire on a chip is given by the following definition - $R = rL/TW$ Where, 'r' is the resistivity of the material 'L' is the length of the Wire 'T' is the thickness of the Wire 'W' is the width of the Wire

Table 31: Resistance, Capacitance and Inductance

Capacitance	Capacitance is the property of a electrical circuit that opposes a change in voltage. It enables a circuit or device to store an electric charge. The unit of measure for Capacitance is Farads. Area capacitance between two wires on a chip is given by the following equation 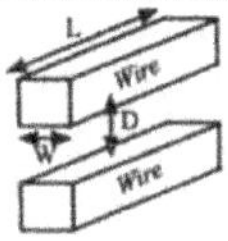 - $C = cp_0LW/D$ Where, *'e' is the dielectric constant* *'p0' is the permittivity of the medium* *'L' is the length of the Wire* *'W' is the width of the Wire* *'D' is the distance between conducting plates*
Inductance	Inductance is the property of a electrical circuit that opposes any change in electric current. The unit of measure for Inductance is Henry.

4. Describe the various terms associated with CMOS Process/Gate?

Various terms associated with CMOS Process/Gate are tabulated below.

Table 32: Various Terms Associated with CMOS Process/Gate

Bulk	The silicon region that a transistor is built in. In the context of HSpice, the bulk is one of the 4 terminals of a transistor. In reality it is not a single node at all - it is fairly resistive silicon which is largely depleted of carriers. The gate oxide is grown on the bulk silicon and transistor source and drain regions are built in the bulk.
Poly	Polysilicon. The material used to make transistor gates. When it runs over field oxide (i.e not over a transistor) it is called 'field poly'. It is silicon but unlike the bulk material, it is deposited, not grown, and it is polycrystalline.

Table 32: Various Terms Associated with CMOS Process/Gate

Gate	Transistor gates are polysilicon deposited on top of a very thin gate oxide. The gate oxide is thermally grown. The polysilicon deposited on the gate oxide is couple of thousand Angstroms thick. The poly gate is doped p-type over p-channel transistors and n-type over n-channel transistors. This is done to help achieve the necessary threshold voltages in each transistor type (the threshold voltage depends (among lot of other things) on the doping density and type of gate). The potential diode problem at the interface of p-type poly and n-type poly is avoided by saliciding the top of the poly. Salicide is a thin layer of Titanium that is deposited on top of he poly and is combined with some of the silicon there to make TiSi2. It is called salicide because it is a 'self-aligned' silicide layer. It is self-aligned because no mask is needed to keep it only on poly and source/drain layers. In addition to shorting out the diode between n-poly and p-poly, the salicide also reduces the resistance of poly by about an order of magnitude, helping to reduce RC delays along gates.
Source/Drain	The Source and Drain are the terminals of a transistor that abut the gate. They are isolated when the gate is off and more or less connected when the gate is on. They are formed by implanting n- or p-type dopants into selected well areas. The source and drain regions need to be highly doped to minimize the contact resistance between the metallization layer and the source/drain region. Source and Drain regions are salicided at the same time as the poly gates. This helps to provide a low-resistance contact to the interconnect above and lowers the resistance of the source/drain regions. If the source/drain regions were not salicided the extra parasitic resistance would rob the transistor of some of its performance. The salicide can be masked off of selected source/drain regions. This is normally done only in I/O buffers. In I/O buffers some series resistance is actually desirable because it helps to attenuate ESD strikes. Drain specifically refers to the terminal of a transistor not connected to the supply and Source specifically refers to the terminal of a transistor connected to the supply. Source and Drain are physically identical.
Channel	A very thin region below the gate of a MOS transistor that can be made conductive by putting an appropriate voltage on the gate of the transistor. It connects the source and drain regions of the transistor.
Isolation	Isolation is what separates transistors from each other. It defines the boundaries of the source/drain regions and also the edges of gates that are not adjacent to source/drain areas. Isolation is important because it prevents one transistor from affecting another and because it provides a nice clean termination for source/drain regions and gate edges. Without such a nice clean termination of these edges we would have excessive leakage along the edge.
Accumulation	When a CMOS gate is turned off beyond the corresponding supply (i.e below VSS for an n-channel or above VCC for a p-channel), the channel area has accumulated carriers of the same type as the bulk doping for that device. For an n-channel device, the bulk is p-type and holes accumulate for voltages less than Vt. For a p-channel device, the bulk is n-type and electrons accumulate for gate voltages greater than (VCC-Vt). These accumulated carriers cannot cause conduction between the source and drain because at least one of these junctions is reverse biased.

Table 32: Various Terms Associated with CMOS Process/Gate

Inversion	When a CMOS gate is turned on (i.e gate voltage is greater than Vt for an n-channel device and below (VCC-Vt) for a p-channel device), a thin layer of minority carriers appears directly under the gate oxide. The type of carriers is opposite that of the channel area doping, hence the term 'Inversion'. In an n-channel device, the channel area is doped p-type and the inversion layer is made up of electrons (hence the term n-channel). Vice-versa for the p-channel device. The carriers in the inversion layer can cause conduction to occur between the Source and Drain terminals because they are of the same type as the carriers in the adjacent Source and Drain regions.
Majority Carriers	Holes in p-type material or electrons in n-type material. The relative number of majority carriers depends on the doping concentration and other stuff.
Minority Carriers	Electrons in p-type material or holes in n-type material. The relative number of minority carriers depends on the doping concentration and other stuff.
Dopant	Extremely low levels of impurity introduced into extremely pure silicon for the purpose of making microprocessors. Pure silicon (and with no light shining on it) is only conductive because of thermally generated carriers (i.e electrons/holes). N-type dopnts (Phosphorus, Arsenic) displace a silicon atom in the crystal lattice and contribute an extra electron to the carrier soup. The n-type dopant atom itself, having contributed its extra electron to the soup, remains as a fixed positive ion stuck in the lattice. P-type dopants (i.e boron) also displace a silicon atom in the lattice but take an electron from the carrier soup (i.e they add a hole). The p-type dopant atom is then negatively charged ion stuck in the lattice. These extra carriers result in increased conductivity.
Field Oxide	This is what separates transistor channels and source/drain regions from each other (except where they are separated by a gate).
ESD	Electrostatic Discharge. This is sort of like a lightning strike on a very small scale. We design I/O buffers to absorb the energy of an ESD strike without being damaged. ESD can also occur during silicon processing due to the use of plasma etches. These etches involve high voltage RF fields that can cause charge to build up on the wafers being processed.
Latch-Up	A phenomenon in which a p-n-p-n device goes into a self sustaining low resistance state. Unfortunately, power and ground are usually connected to the ends of the pnpn device and the resulting supply current flow is catastrophic. Latch-Up can be induced by a variety of means both in the I/O circuits of a processor and in strictly internal circuits. Latch-Up is avoided by the use of epitaxial wafers and by good layout practices. Latchup is generally worse at high operating temperatures because the resistance of silicon is higher and CMOS latchup is caused by IR drops in silicon.
Miller Capacitance	This is the capacitance between a transistor gate and the source and drain regions. Part of this capacitance is because the source and drain regions extend under the gate poly and part is due to the fringing capacitance between the side of the gate poly and the outlying source/drain region. In practice, Miller capacitance often refers to the total capacitance between a logic gate's output and its input.

Table 32: Various Terms Associated with CMOS Process/Gate

Miller Effect	The phenomenon that a logic gate's switching output affects the input of that gate due to the capacitance between the gate and the output. Some of the capacitance is between the poly gate and the drains of the output. If one or more of the logic gate's transistors are turned on then there is additional capacitance between the channel and the gate. The output signal is capacitively coupled to the input, making the input of the gate appear more capacitive than it would otherwise be. Because of the additional capacitance between the channel and the gate terminal, transistors that are turned on exhibit this effect most.

5. What are the most common guidelines used to ensure Circuit Robustness?

Following are some of the most common guidelines used to ensure Circuit Robustness

Table 33: Guidelines to Ensure Circuit Robustness

1	Use complementary circuits as they have good noise margins.
2	Do not use pseudo nMOS as they have very low noise margin and has steady current flow which increases power.
3	Use synchronous logic where ever possible.
4	Limit stack heights in complementary logic gates to 3 or less for PMOS and 4 or less for NMOS (these are typical values used in the industry for nano technology process).
5	Nodes that are prone to cross talk, leakage etc. should be observed during layout to ensure minimization of these phenomena on the final layout.

6. What do you mean by effect of Noise in a design and what are the most common techniques used to reduce its effect?

Noise is any phenomenon that causes the voltage of a signal net to deviate from the nominal supply or ground voltage for reasons other than switching of the gate that drives the signal. The net that has noise induced on it is called the Victim net and the net that induces noise on its neighbors is called the Aggressor net. Noise can be classified into 4 categories as shown in Table below depending on the direction in which it causes the deviation.

Table 34: Noise Categories

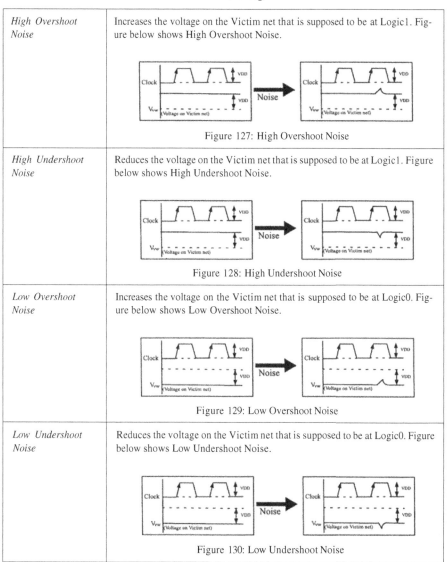

High Overshoot Noise	Increases the voltage on the Victim net that is supposed to be at Logic1. Figure below shows High Overshoot Noise.
	Figure 127: High Overshoot Noise
High Undershoot Noise	Reduces the voltage on the Victim net that is supposed to be at Logic1. Figure below shows High Undershoot Noise.
	Figure 128: High Undershoot Noise
Low Overshoot Noise	Increases the voltage on the Victim net that is supposed to be at Logic0. Figure below shows Low Overshoot Noise.
	Figure 129: Low Overshoot Noise
Low Undershoot Noise	Reduces the voltage on the Victim net that is supposed to be at Logic0. Figure below shows Low Undershoot Noise.
	Figure 130: Low Undershoot Noise

Table below lists the most common sources of Noise.

Table 35: Noise Sources with Description

Noise Source	Description
Capacitive Coupling Noise	This is the most common source of noise in deep submicron designs. Capacitive Coupling noise is the noise pulse as a result of capacitance between two neighboring wires. There always exists a capacitance between any two neighboring wires. As a result of this capacitance between neighboring wires, a transition on one wire (aggressor) causes a voltage pulse (the strength of this pulse is proportional to C*dV/dt (where 'C' is the side wall capacitance and 'dV/dt' is the rate of change of voltage on the aggressor)) on its neighbor (victim). This is because of the electric field that exists across the capacitor. This voltage pulse causes either a overshoot or an undershoot on the voltage level on the victim net which in some cases might result in a functional failure. Figures 131, 132 and 133 below shows few cases where noise due to capacitive coupling results in a functional failure. Figure below shows functional failure as a result of capacitive coupling between net 'd' and net 'a'. In the figure because of capacitive coupling between net 'd' and net 'a', net 'd' throws a voltage pulse on net 'a' whenever it makes a transition from logic 1 to logic0 or logic0 to logic1. In the figure we see that when net 'd' makes a transition from logic 1 to logic0 during Phase A of Clock it results in a low undershoot on net 'a' and since the transition on net 'd' happens somewhere in the middle of Phase A, net 'a' has enough time to dissipate the undershoot as a result of which correct value gets latched as can be seen from the value on net 'c' during Phase B of Clock. Also in the figure when net 'd' makes a transition from logic0 to logic 1 sometime during Phase C of Clock it results in a low overshoot on net 'a', but since the transition on net 'd' happens just before Phase D of Clock, net 'a' doesn't have enough time during Phase C of Clock to dissipate the overshoot as a result of which incorrect value gets latched as can be seen from the value on net 'c' during Phase D of Clock. 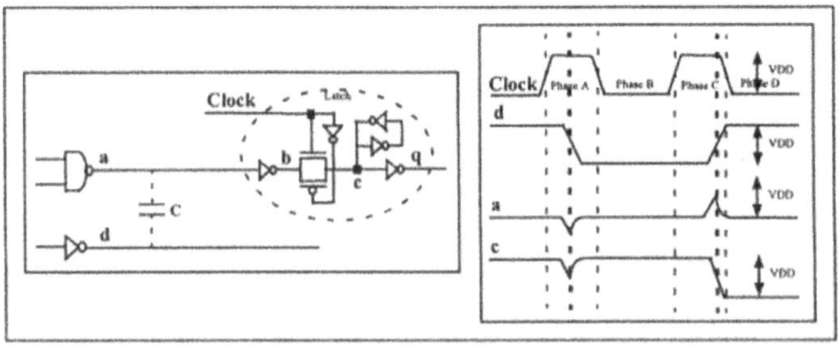 Figure 131: Capacitive Coupling between Two Signal Nets Figure below shows functional failure as a result of capacitive coupling between net 'd' and Clock net. In the figure because of capacitive coupling between net 'd' and Clock net, net 'd' throws a voltage pulse on Clock net whenever it makes a transition from logic 1 to logic0 or logic0 to logic1. In the figure we see that when net 'd' makes a transition from logic0 to logic 1 sometime during Phase B of clock it results

Table 35: Noise Sources with Description

Noise Source	Description
	in a low overshoot on Clock net as a result of which the latched value in Phase B gets overwritten with a new incorrect value as can be seen by the value on net 'c' during Phase B of clock. 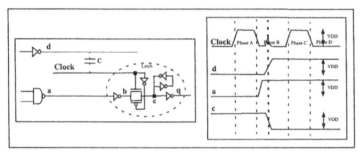 Figure 132: Capacitive Coupling between a Signal Net and a Clock Net Figure below shows functional failure due to capacitive coupling in the path with dynamic gates. In the figure because of capacitive coupling between net 'a' and net 'd', net 'd' throws a voltage pulse on net 'a' whenever it makes a transition from logic 1 to logic0 or logic0 to logic1. In the figure we see that when net 'd' makes a transition from logic0 to logic 1 during Phase A of Clock it results in a low overshoot on net 'a' as a result of which incorrect value gets latched as can be seen from the value on net 'g' during Phase B of clock. 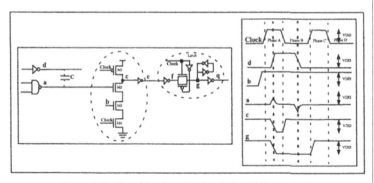 Figure 133: Capacitive Coupling in a Path with Dynamic Gates Figure below shows the dominance of Coupling Capacitance over other Capacitances associated with an Interconnect as we move towards smaller geometries.

Table 35: Noise Sources with Description

Noise Source	Description
	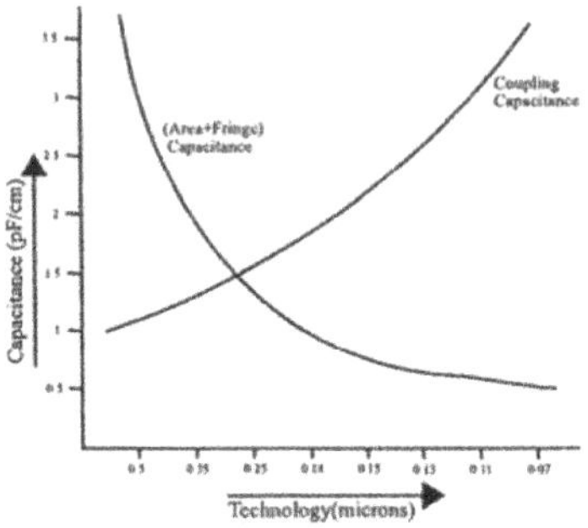 Figure 134: Coupling Capacitance Over Other Capacitances with Technology
Inductive Coupling Noise	This is caused by magnetic fields induced by currents flowing through neighboring wires. This has become of increasing concern in deep submicron technology as the low permittivity materials that act to reduce capacitance have increased permeability and are therefore more susceptible to inductive coupling. The failures caused in circuits by inductive coupling is similar to the ones caused by capacitive coupling discussed above.
Leakage Noise	Main contributors to this are 1. Leakage currents through diode junctions 2. Subthreshold conduction through transistors (this is the current flowing through transistors even when they are not conducting). Subthreshold conduction occurs when a transistor has a V_{gs} greater than zero (in case of NMOS transistors) but less than V_t. In this region the transistor will exhibit a drain current which is exponentially dependent on both V_{gs} and V_{ds}. Although the magnitude of this current is fairly small, it is non zero and especially in the case of dynamic circuits it can be a significant contributor to overall noise margin. In the figure below while the inputs 'a' and 'b' in the 2-input domino NAND gate are low during the evaluate phase of the clock, due to sub-threshold leakage current through transistors M1 and M2 we may result in draining away the charge on node 'c' leading to degradation in its voltage level and a wrong value at the output node 'd'. 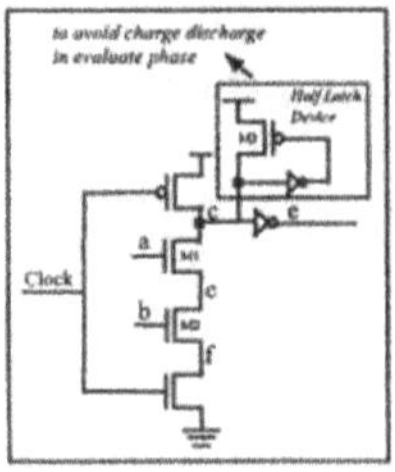 Figure 135: 2-input Domino NAND Gate

Table 35: Noise Sources with Description

Noise Source	Description
	In the above leakage noise due to subthreshold conduction is avoided by having a Half Latch device. 3. Leakage through gate of a transistor The primary contributing factor for this phenomenon is local variations in the power supply voltage.
Charge Sharing Noise	This is the noise induced at a dynamic node due to charge sharing (charge redistribution) between the dynamic node and some of the internal nodes of the gate. It is of a prime concern in dynamic gates. To illustrate this let us consider a 2-input NAND gate as shown in the figure below. From the waveform we see that during the first evaluate phase when inputs 'a' and 'b' arc logic1, both nodes 'e' and 'c' are discharged. In the following precharge phase let us assume that input 'a' is logic0 as a result of which node 'c' will be prechargcd by transistor M1 and node 'e' will remain at logic0. In the next evaluate phase if 'a' is logic1 and 'b' is logic0 then there will be charge sharing between nodes 'c' and 'e' as a result of which node 'e' will be pulled high and node 'c' will be pulled low. If the voltage on node 'c' is reduced by a large amount then node 'd' may switch causing node 'd' to be wrongly set to logic1. Figure 136: Charge Sharing Noise In the figure above charge sharing is avoided by precharging the internal node in the NMOS tree (i.e node 'c') during the precharge phase of the clock.
Power Supply Noise	This refers to noise on the Power (P) and Ground (G) nets that gets passed on to the signal nets by the transistors connected to these PG nets. The components of noise on PG nets are - 1. IR drop on each element of PG net because of the finite resistance offered by these elements. 2. RLC response of the chip and package to current demands that peak at the beginning of a clock cycle.

Table below lists the techniques used to reduce the effect of noise sources mentioned in the table above.

Table 36: Techniques to Reduce Noise

Noise Source	Techniques to Reduce Noise
Capacitive Coupling Noise	**1.** Staggering gates reduces noise due to capacitive coupling. This is because of two reasons - a. Since the length of overlap between adjacent wires is reduced the effective coupling capacitance between the adjacent wires (i.e Victim and Aggressor) is reduced (as capacitance is directly proportional to the length of overlap (i.e $C=pLW/D$)) which results in a reduced noise on the Victim as can be seen from figure below. b. The overshoot noise on Victim (i.e net1 in figure below) as a result of staggered gate on the Aggressor tend to cancel itself with the undershoot noise on Victim (i.e net1) as a result of which the effective noise on Victim (i.e net1) because of capacitive coupling between net1 and net3/net4 is reduced as can be seen from figure below. Eventhough the cancellation is not perfect it is still effective. 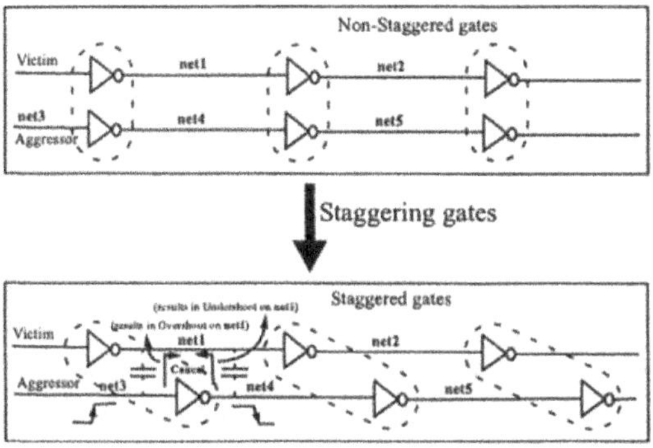 Figure 137: Staggering Gates **2.** Shielding Victim by manually running supply/ground wire on one or both sides of the Victim or above or below the Victim in the same or a different metal layer reduces noise due to capacitive coupling. This is because of the reduced value of **dV/dt** (since aggressor in this case would be VDD/VSS, **dV/dt**=0) in the equation for strength of noise pulse on the Victim (i.e strength of noise pulse on the Victim = (Coupling capacitance) x (Rate of change of voltage on the Aggressor)). This can be seen from figure below.

Table 36: Techniques to Reduce Noise

Noise Source	Techniques to Reduce Noise
	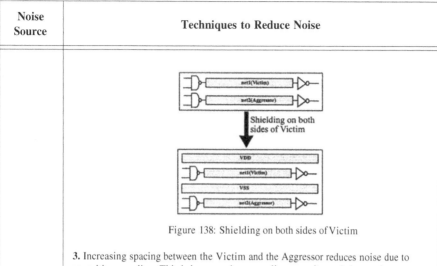 Figure 138: Shielding on both sides of Victim **3.** Increasing spacing between the Victim and the Aggressor reduces noise due to capacitive coupling. This is because since coupling capacitance is inversely proportional to the distance between Victim and Aggressor (i.e $C=pLW/D$), the increase in spacing between Victim and Aggressor reduces the value of coupling capacitance between them thereby reducing the noise due to coupling capacitance. This can be seen from figure below. 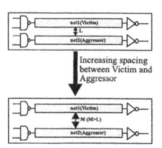 Figure 139: Increasing Spacing between Victim and Aggressor **4.** Increasing the width of Victim net reduces noise due to capacitive coupling. This is because increasing width of Victim net offers less resistance to Victim driver (since resistance of a net is inversely proportional to its width) as a result of which Victim driver results in dissipating the noise voltage at a much faster pace than before because of its improved drive strength (i.e since it sees less loading). This can be seen from figure below.

Table 36: Techniques to Reduce Noise

Noise Source	Techniques to Reduce Noise
	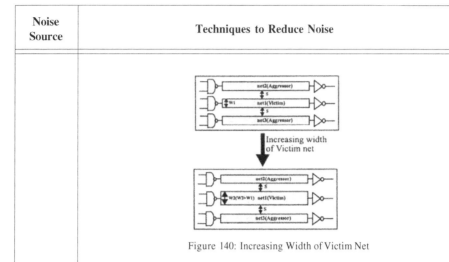 Figure 140: Increasing Width of Victim Net **5.** Inserting buffer on very long victim nets reduces noise due to capacitive coupling. This is because inserting buffer reduces the length of the net and redistributes capacitive coupling between two newly created nets as shown in figure below. Since coupling capacitance is directly proportional to the length of Victim overlap, the smaller the length the smaller the coupling capacitance as a result of which we see less noise due to coupling capacitance. Also adding buffer improves the signal strength on the Victim nets (i.e net1_0, net1_1 see better signal strength as drivers driving these nets see less loading) which helps in faster dissipation of any noise voltage induced on them. 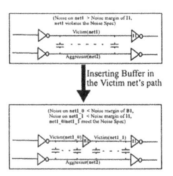 Figure 141: Inserting Buffer in the Victim Net's Path **6.** When both the Aggressors and Victim are on the same metal layer moving the Aggressors on to a different metal layer reduces the effect of capacitive coupling noise. This is because when we move the Aggressors to a different metal layer the area of overlap between the Victim and Aggressor goes down whereas the distance between them goes up. Since capacitance is directly proportional to area of overlap and inversely proportional to distance the value of coupling capacitance goes down as a result of which we see less noise on Victim because of coupling capacitance.

Table 36: Techniques to Reduce Noise

Noise Source	Techniques to Reduce Noise
	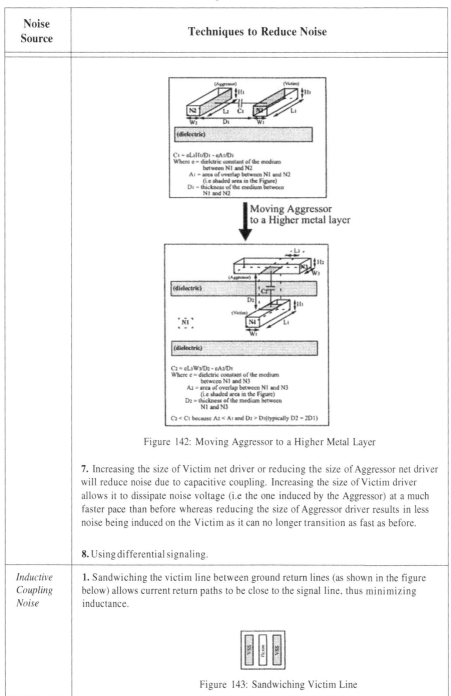 Figure 142: Moving Aggressor to a Higher Metal Layer

7. Increasing the size of Victim net driver or reducing the size of Aggressor net driver will reduce noise due to capacitive coupling. Increasing the size of Victim driver allows it to dissipate noise voltage (i.e the one induced by the Aggressor) at a much faster pace than before whereas reducing the size of Aggressor driver results in less noise being induced on the Victim as it can no longer transition as fast as before.

8. Using differential signaling.

Inductive Coupling Noise	**1.** Sandwiching the victim line between ground return lines (as shown in the figure below) allows current return paths to be close to the signal line, thus minimizing inductance. Figure 143: Sandwiching Victim Line

Table 36: Techniques to Reduce Noise

Noise Source	Techniques to Reduce Noise
	2. Having dedicated ground planes in the layers above and below the victim line (as shown in the figure below) will result in reducing inductance as the ground planes provides an excellent return paths for the signal current. 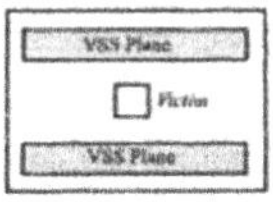 Figure 144: Dedicated Ground Planes **3.** Splitting wider victim wires into thinner wires with shields in between (as shown in figure below) results in reducing self-inductance. 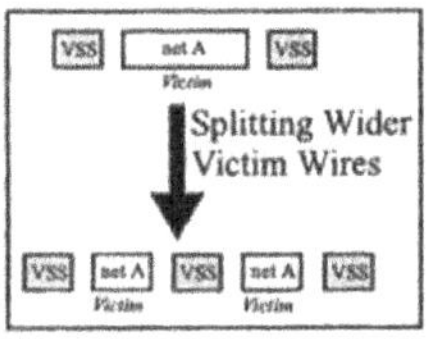 Figure 145: Splitting Wider Wires into Thinner Wires with Shields in between **4.** Using staggered inverter patterns (as shown in the figure below) results in reducing inductance effects. 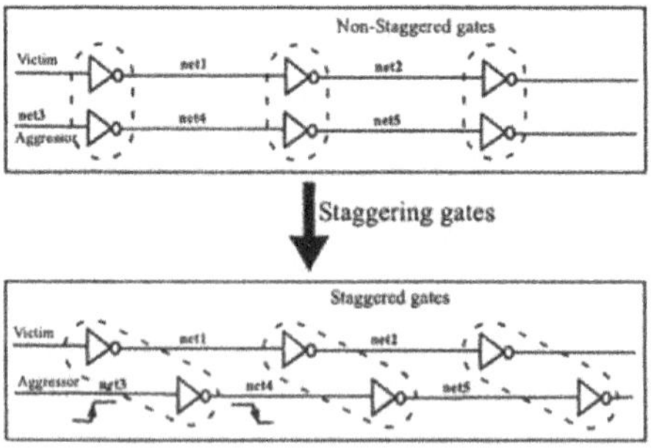 Figure 146: Staggered Inverter Patterns **5.** Using a twisted bundle layout structure results in minimizing inductive coupling noise as they create complementary and opposite current loops in the layout structure such that the magnetic fluxes arising from any signal net within a twisted group cancel each other in the current loop of a net of interest.
Leakage Noise	**1.** Use high Vt gates in the regions where we see increased effects of leakage noise. **2.** Using Half Latch device on dynamic nodes (as shown in Figure 135).

Table 36: Techniques to Reduce Noise

Noise Source	Techniques to Reduce Noise
Charge Sharing Noise	**1.** Using anti-charge sharing device on internal nodes (as shown in Figure 136).
Power Supply Noise	**1.** Make the Power and Ground nets Wider and Denser in the regions where your IR drop does not the meet the IR Spec. **2.** Add on-chip Decoupling capacitors in the regions where there is demand for huge currents.

7. What is Electromigration and IR drop?

Electromigration

If current density in any element of the power grid or signal exceeds the process limits then the element could fail (break) due to overheating or migration of metal ions under electrical field or a combination of both. This phenomenon is known as Electromigration. Table below shows the most common techniques used to reduce the effect of Electromigration in a design:

Table 37: Techniques to Reduce the Effect of Electromigration

1	Metal Slotting.
2	Increasing the Width of Metal layer.
3	Adding more Vias.

IR drop

The finite resistance of power grid generates a drop in voltage across each element resulting in reduced supply voltage at the transistor. This reduced voltage could effect timing and hence is controlled by a budget. This phenomenon is known as IR drop. Table below shows the most common techniques used to reduce IR drop in a design.

Table 38: Techniques to Reduce IR drop

1	Increasing the Width of Metal layer (i.e Power layer).
2	Adding more Power and Ground layers near the region experiencing IR drop.
3	Adding more Vias.

8. What are Differential Sense Amplifiers?

Differential Sense Amplifiers (DSA) are circuits used by high performance memory designs to achieve fast access time. The need to have the Cache memory cell as small as possible (in order to have large cache sizes on chip), results in a poorer drive strength for the cell and makes it impossible to swing its large output load (bitline capacitance) full rail within a reasonable time. For this reason Caches and other memory structures use differential sense amplifiers which sample a small signal swing of the bit line discharge and amplifies it to a full rail output, thereby speeding up memory access. Since sense amplifier is not a standard static CMOS circuit, extreme care must be taken in both designing the circuit and the final layout. The two most common differential sense amplifier circuits used in the industry are gate fed differential sense amplifier and drain fed differential sense amplifier. In the gate fed DSA the input to the sense amplifier is fed to the gates where as in the drain fed the inputs are fed to the drain of the sense amplifier. One of the advantages of gate fed DSA over drain fed DSA is having reduced bit loading as the inputs are fed to the gates, and the isolation of the sense amplifier nodes from the bit lines provides the feasibility of having larger drive capabilities.

9. What is Antenna Effect and what are the most common techniques used to reduce it in a design?

Antenna effect is a phenomenon of transistor gate oxide damage as a result of charge buildup on the floating conductors during one of the processing steps (i.e plasma etching, ion implantation or photoresist strip). If the gate of a transistor is connected to a metal interconnect having a large area then during etching of the metal, the metal area acts as an antenna collecting ions and rising in potential. When the rising voltage on the gate (i.e because of the rising potential on the interconnect to the gate) reaches a point where it is equal to the gate oxide breakdown voltage then the gate oxide breaks resulting in a very low resistance path between the gate and the channel. In order to avoid this from happening every manufacturer provides certain antenna rules to be followed while building the masks. Antenna rules specify the limits of the maximum ratio of interconnect area to connected active poly gate area in cases where the interconnect at some stage during processing is not connected to diffusion (i.e p/n source or drain) i.e if a gate does not get connected to diffusion until Metal 5 deposition then any segments of the floating network on poly interconnect, Metal1, Metal2, Metal3 or Metal4 each must not exceed the maximum area ratio of that segment to total active poly area. These rules ensure that processing steps (i.e plasma etching, ion implantation or photoresist strip) which can result in charge collection do not result in transistor gate oxide damage due to excessive charge collection when the metal is floating (i.e gates not yet tied to source/drain diffusion).

Table below shows the most common techniques used to reduce/eliminate Antenna effect in a design.

Table 39: Techniques to Reduce Antenna Effect

1	Inserting jumpers breaks up a long wire (i.e segment of a wire resulting in antenna effect) so that the segment of the wire connected to the gate input is shorter thereby collecting less charge and thus eliminating antenna effect. The disadvantage of using this technique is that it causes routing congestion problems of upper metal layers. Figure below shows the way Jumpers are used to eliminate antenna effect.

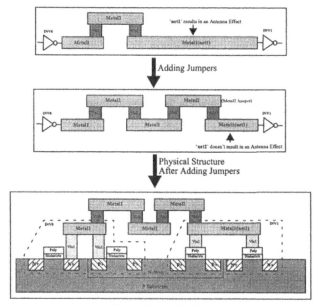

Figure 147: Adding Jumpers

Table 39: Techniques to Reduce Antenna Effect

2	Inserting diode at the gate input provides a conduction path to the substrate so that the built up charges can be directed to the substrate/well before it damages the transistor gate oxide. The disadvantage of using this technique is that it results in increase in area and timing (as 'net1' in figure below sees more loading). Figure below shows the way Diodes are used to eliminate antenna effect.

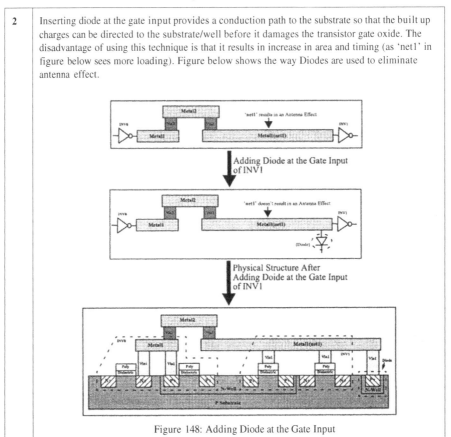

Figure 148: Adding Diode at the Gate Input

10. What do you mean by a Simulation Corner and what are the typical simulation corners used for Standard Cell Characterization, Max time, Min time, Electromigration, IR drop, Noise and Power?

Simulation Corner is the Transistor model (slow, typical, fast), Wire model (slow, typical, fast), Temperature (low, nominal, high, burn-in) and Voltage (ultra low, low, nominal, high, burn-in) used to simulate a circuit. They are usually annotated by a set of five letters that represent the Transistor models (i.e P and N), Supply Voltage, Temperature and Wire model. Table below shows the typical simulation corners used for Standard Cell Characterization, Max time, Min time, Electromigration, IR drop, Noise and Power in the industry.

Table 40: Typical Simulation Corners Used for Various Simulations

Type of Simulation	Simulation Corners Used
Standard Cell Characterization	TTLH-T (Typical P-transistor Model, Typical N-transistor Model, Low Voltage, High Temperature, Typical Wire Model)
Max time	TTLH-T (Typical P-transistor Model, Typical N-transistor Model, Low Voltage, High Temperature, Typical Wire Model)
Min time	FFHL-F (Fast P-transistor Model, Fast N-transistor Model, High Voltage, Low Temperature, Fast Wire Model)
Electromigration	FFHH-F (Fast P-transistor Model, Fast N-transistor Model, High Voltage, High Temperature, Fast Wire Model)
IR Drop	TTNH-T (Typical P-transistor Model, Typical N-transistor Model, Nominal Voltage, High Temperature, Typical Wire Model)
Noise	FFBB-S (Fast P-transistor Model, Fast N-transistor Model, Burn-in Voltage, Burn-in Temperature, Slow Wire Model)
Power	TTNH-T (Typical P-transistor Model, Typical N-transistor Model, Nominal Voltage, High Temperature, Typical Wire Model)

11. What are the various types of Pass Gate Mux Library elements used in a CPU design?

Table below shows the various types of Pass Gate Mux Library elements used in a CPU design.

Table 41: Types of Pass Gate Mux Library Elements

Type of Pass Gate Mux	Library Cell	Typical Usage
One-Hot Inverting Pass Gate Mux	*mux2i* (2-input one-hot inverting pass gate Mux) 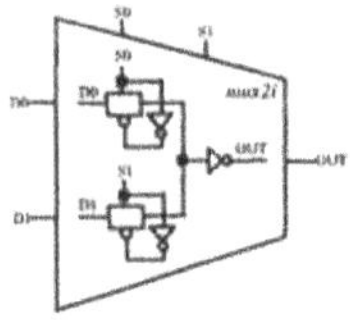	If you are determined to use a pass gate Mux then use one-hot inverting pass gate Mux where the following 3 conditions are true - 1. In cases where a Mux is used as one of the gates in a path within the block, using this type of Mux gives the best possible delay number through the data inputs (i.e D0 and D1). 2. In cases where more than one Mux gets the same selects, it is good to have decode logic for the selects outside the Mux (as is the case here) as it results in a more area efficient design when compared to encoded version. 3. You are aware of the signal contention issue in scan and functional mode and you are planning on taking care of it outside the Mux.
One-Hot Non-Inverting Pass Gate Mux	*mux2* (2-input one-hot non-inverting pass gate Mux) 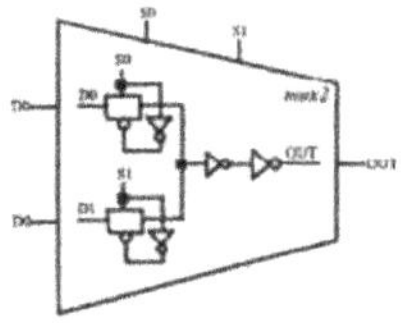	If you are determined to use a pass gate Mux then use one-hot inverting pass gate Mux where the following 3 conditions are true - 1. In cases where a Mux is used to drive a signal out of the block, using this type of Mux provides good drive strength. 2. In cases where more than one Mux gets the same selects, it is good to have decode logic for the selects outside the Mux (as is the case here) as it results in a more area efficient design when compared to encoded version. 3. You are aware of the signal contention issue in scan and functional mode and you are planning on taking care of it outside the Mux.

Table 41: Types of Pass Gate Mux Library Elements

Type of Pass Gate Mux	Library Cell	Typical Usage
One-Hot Buffered Pass Gate Mux	*mux2b* (2-input one-hot buffered pass gate Mux) 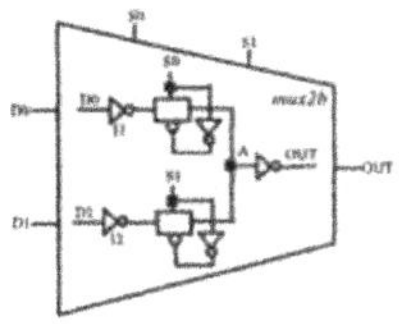	If you are determined to use a pass gate Mux then use one-hot inverting pass gate Mux where the following 3 conditions are true - 1. In cases where the inputs coming from a distant block feeds directly into a Mux, using this type of Mux gives a better slew rate at the internal node (i.e node A in figure) thereby giving a better signal transition at the output of the Mux (i.e node OUT in figure). Also here we don't have to worry about any charge sharing issues as the input buffer (i.e I1 and I2) isolates the huge wire capacitance outside the block and the pass gate internal to the Mux. 2. In cases where more than one Mux gets the same selects, it is good to have decode logic for the selects outside the Mux (as is the case here) as it results in a more area efficient design when compared to encoded version. 3. You are aware of the signal contention issue in scan and functional mode and you are planning on taking care of it outside the Mux.
Encoded Inverting Pass Gate Mux	*mux4ei* (4-input encoded inverting pass gate Mux) 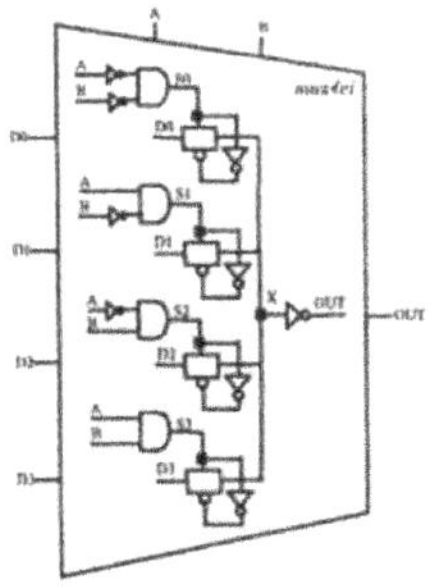	If you are determined to use a pass gate Mux then use encoded inverting pass gate Mux where the following 2 conditions are true - 1. In cases where a Mux is used as one of the gates in a path within the block, using this type of Mux gives the best possible delay number through the data inputs (i.e D0 and D1). 2. In cases where you don't want the additional burden of worrying about signal contention issue in scan and functional mode.

Table 41: Types of Pass Gate Mux Library Elements

Type of Pass Gate Mux	Library Cell	Typical Usage
Encoded Non-Inverting Pass Gate Mux	*mux4e*(4-input encoded non-inverting pass gate Mux)	If you are determined to use a pass gate Mux then use encoded non-inverting pass gate Mux where the following 2 conditions are true - 1. In cases where a Mux is used to drive a signal out of the block, using this type of Mux provides good drive strength. 2. In cases where you don't want the additional burder of worrying about signal contention issue in scan and functional mode.
Encoded Buffered Pass Gate Mux	*mux4eb* (4-input encoded buffered pass gate Mux)	If you are determined to use a pass gate Mux then use encoded buffered pass gate Mux where the following 2 conditions are true - 1. In cases where the inputs coming from a distant block feeds directly into a Mux, using this type of Mux gives a better slew rate at the internal node (i.e node A in figure) thereby giving a better signal transition at the output of the Mux (i.e node OUT in figure). Also here we don't have to worry about any charge sharing issues as the buffer isolates the huge wire capacitance outside the block and the pass gate internal to the Mux. 2. In cases where you don't want the additional burder of worrying about signal contention issue in scan and functional mode.

12. What is the importance of adding Metal Fill Patterns in a Layout?

CMP (Chemical Mechanical Polishing) has emerged as the primary semiconductor fabrication process for planarizing interlayer dielectrics but it is hampered by its sensitivity to layout patterns which cause certain regions on a chip to have thicker dielectric layers than other regions (reason being CMP process tends to polish dielectric over isolated features much more rapidly than that over dense areas). This interlayer dielectric (ILD) variation must be kept in control as this could potentially reduce yield and impact circuit performance. Metal fill patterning has been used to reduce ILD variation due to its sensitivity to layout patterns. Metal fill patterning is a technique of filling large open areas on each metal layer with a metal pattern, which is either grounded or left floating to compensate for pattern dependent ILD thickness variation. Adding Metal Fill results in maintaining a uniform metal density across the chip on each layer thereby ensuring planarity during CMP process. Foundries today typically impose metal pattern density rule which in most cases is "within any window of size YxY (i.e typically 200u x200u for 0.09u Process) the metal pattern density should be greater than or equal to 30%". Figure below shows the effect on ILD thickness variation by adding Metal Fill patterns in the layout.

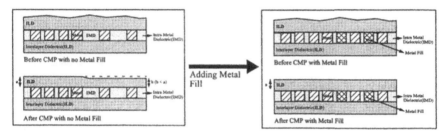

Figure 149: Effect on ILD thickness variation by Addition of Metal Fill Pattern

Care should be taken while adding metal fills as blindly adding metal fill could result in an increased delay on a net (say net1) because of capacitive coupling between the added metal fill and its neighbour (net1 in this case). The advantages and disadvantages of using grounded and floating metal fills are tabulated below.

Table 42: Grounded against Floating Metal Fill

Fill Technique	Advantages	Disadvantages
Grounded Metal Fill	1. Since they are at a known potential the noise voltage it offers because of its coupling with its neighbors is very low.	1. Additional burden of connecting all the metal fills to ground.
Floating Metal Fill	1. No additional burden of connecting the metal fills to ground.	1. Noise voltage it offers because of its coupling with its neighbor is more than that offered by a Grounded Metal Fill.

13. What are the most common Layout Schemes used for implementing a Power Network?

The three most common Layout Schemes for implementing a Power network are Power Plane, Power Grid and as a Routed Network. Table below summarizes their characteristics.

Table 43: Power Network Implementation Schemes

Char ester is tics	Power Plane	Power Grid	Routed Network
Implementation	Figure below shows a typical Power Plane implementation of a Power Network.	Figure below shows a typical Power Grid implementation of a Power Network.	Figure below shows a typical Routed Network implementation of a Power Network.
	Figure 150: Power Plane	Figure 151: Power Grid	Figure 152: Power Network
IR Drop	Very Low	Low	High
Inductive Drop	Low	High	Very High
Electromigration	Very Low	Low	High Probability
Usage	Will be common for High frequency designs (i.e GHz designs)	Most popular	Low cost systems

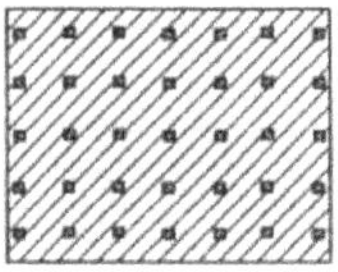

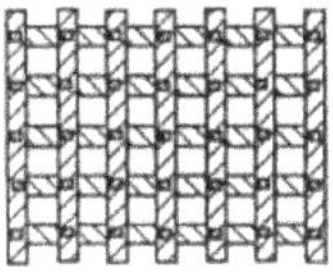

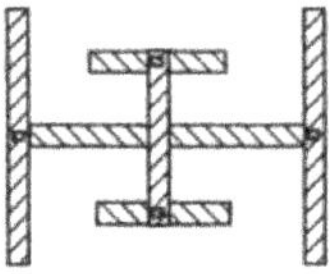

4 Verification and Testing

1. What are some of the responsibilities of a Verification Engineer in the Chip Industry?

Some of the responsibilities of Verification Engineer are summarized below.

Table 44: Verification Engineer Responsibilities

1	Develop Stand Alone Test Bench environment for a block.
2	Develop Full Chip regression environment for the Chip.
3	Write Directed diags for the design.
4	Write Functional Coverage Objects for various blocks within the Chip, run the simulation with these coverage objects in place, generate reports and use the reports to generate additional tests to cover the functionality.
5	Run Formal Functional verification tools to find bugs within the Design.
6	Model turnin, Job submission and Tracking of results.
7	Configuration management administration.
8	Develop Pseudo Random Test generator.
9	Generate Random tests (weighted or templatized) using Pseudo Random Test generator.
10	Debug test failures.
11	Come up with a verification test plan (i.e things he want to test, how he want to test (i.e self-checking diags or simulating against a reference model (instruction accurate architectural model) and comparing the architectural state or through assertions or eye boiling the simulation output etc.)
12	Filing bugs through a bug tool and keeping track of it until they get resolved.
13	Providing play back vectors for post-silicon debug (i.e for tester).
14	Writing Monitors, Checkers or Assertions for better functional coverage and early detection of bugs in the simulation environment.
15	Running Code Coverage tools and generating reports for various types of code coverages (i.e Path, Block, Expression, Toggle, State etc.) and using the reports to generate additional tests to cover remaining portion of the coverage.

2. What are some of the responsibilities of a Test Engineer in the Chip Industry?

Some of the responsibilities of Test Engineer are summarized below.

Table 45: Test Engineer Responsibilities

1	Running ATPG test patterns.
2	Provide architectural definition for TAP (Test Access Port) controller for the Chip.
3	Provide architectural definition for any BIST controllers within the Chip.
4	Define tester interface for the Chip.
5	Work on post silicon debug.
6	Work with the process guys to resolve any process related post silicon issues.
7	Come up with a list of debug features he would like to have.

3. What are the Verification metrics used for Tapeout?

Table below lists the Verification metrics used for Tapeout.

Table 46: Verification Metrics for Tapeout

1	Bug rate in the last few weeks before tapeout is zero.
2	Design went through more than 10 billion cycles of simulation without a bug.
3	Code and Functional coverage close to 100%.

4. What is a Monitor, Checker and Assertion?

Table below provides definitions for Monitors, Checkers and Assertions.

Table 47: Monitor, Checker and Assertion

	Definition
Monitor	It displays information about the design or test while the test is running. They are inserted anywhere in the simulation environment or design that requires monitoring. In many cases designer or verification engineer may want to be notified when a particular event or a series of events happen in a simulation. Simulation may flag the condition with a simple print statement. A simple Monitor (verilog code) observing signals in decode stage of a pipeline is shown below -

```
'timescale 1ns/1ns

module inst_monitor(ifu_i0, ifu_i1, ifu_i0_vld, ifu_i1_vld, i0_rs1_vld, i0_rs2_vld, i0_rd_vld,
i1_rs1_vld, i1_rs2_vld, i1_rd_vld, clk);

input [31:0] ifu_i0, ifu_i1;
input      ifu_i0_vld, ifu_i1_vld;
input      i0_rs1_vld, i0_rs2_vld, i0_rd_vld;
input      i1_rs1_vld, i1_rs2_vld, i1_rd_vld, clk;

always @(posedge clk)
  begin
  $display("*************************** @ %0dns ***************************", $time);
  #2
  case({ifu_i1_vld, ifu_i0_vld})
  2'b00 : $display("No valid instructions this cycle");
  2'b01 : begin
            $display("i0 is valid, i1 is invalid");
            if(ifu_i0[31:30] == 2'b11) $display("i0 is LOAD");
            else if(ifu_i0[31:30] == 2'b10) $display("i0 is ADD");
            else if(ifu_i0[31:30] == 2'b01) $display("i0 is CALL");
            else $display("i0 is BRANCH");
            $display("D Stage i0_rs1_vld=%b, i0_rs2_vld=%b, i0_rd_vld=%b", i0_rs1_vld, i0_rs2_vld, i0_rd_vld);
          end
  2'b10 : begin
            $display("i0 is invalid, i1 is valid");
            if(ifu_i1[31:30] == 2'b11) $display("i1 is LOAD");
            else if(ifu_i1[31:30] == 2'b10) $display("i1 is ADD");
            else if(ifu_i1[31:30] == 2'b01) $display("i1 is CALL");
            else $display("i1 is BRANCH");
            $display("D Stage i1_rs1_vld=%b, i1_rs2_vld=%b, i1_rd_vld=%b", i1_rs1_vld, i1_rs2_vld, i1_rd_vld);
          end
  2'b11 : begin
            $display("Both i0 and i1 are valid");
            if(ifu_i0[31:30] == 2'b11) $display("i0 is LOAD");
            else if(ifu_i0[31:30] == 2'b10) $display("i0 is ADD");
            else if(ifu_i0[31:30] == 2'b01) $display("i0 is CALL");
            else $display("i0 is BRANCH");
            $display("D Stage i0_rs1_vld=%b, i0_rs2_vld=%b, i0_rd_vld=%b", i0_rs1_vld, i0_rs2_vld, i0_rd_vld);
            $display("\n");
            if(ifu_i1[31:30] == 2'b11) $display("i1 is LOAD");
            else if(ifu_i1[31:30] == 2'b10) $display("i1 is ADD");
            else if(ifu_i1[31:30] == 2'b01) $display("i1 is CALL");
            else $display("i1 is BRANCH");
            $display("D Stage i1_rs1_vld=%b, i1_rs2_vld=%b, i1_rd_vld=%b", i1_rs1_vld, i1_rs2_vld, i1_rd_vld);
          end
  default : $display("No valid instructions this cycle");
  endcase
  $display("\n \n");
  end

endmodule
```

Table 47: Monitor, Checker and Assertion

	Definition
Checker	In addition to having the properties of a Monitor, Checkers check for correctness of the design while the test is running. They allow for quick detection and correction of bugs at the source, rather than relying on those failures being propagated through the test bench. A simple Checker (verilog code) checking for correctness of Instruction ID generation logic (instruction ID generally gets used by units down the pipe (i.e units after Decode Unit) to index one of their queue structures or to determine the age of an instruction) is shown below - 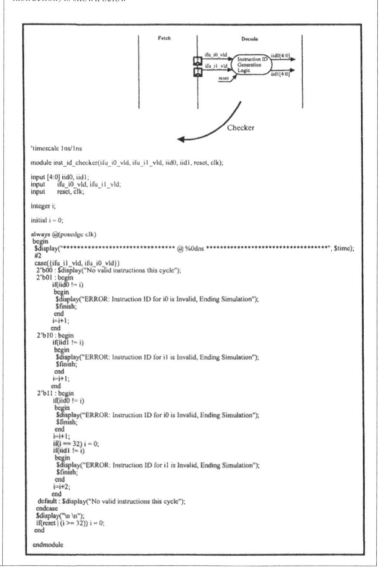

Table 47: Monitor, Checker and Assertion

	Definition
Assertion	Assertion is a statement of design intent, or an assumption about a particular logic behavior, or the behavior of an interface written in assertion language (0-in, System Verilog, OVL etc.) normally placed in line with the RTL code in the comments section to help Simulation and Formal verification tools in finding bugs in the design. Assertions require comprehensive set of stimuli from a simulation environment in order to exercise them. Formal verification tools exhaustively exercise assertions by proving or disproving them without the use of external stimuli. Like Checkers, Assertions allow for quick detection and correction of bugs at the source, rather than relying on those failures being propagated through the test bench. Assertion is synonymous with Property. A simple Assertion (0-in assertion) is shown below - 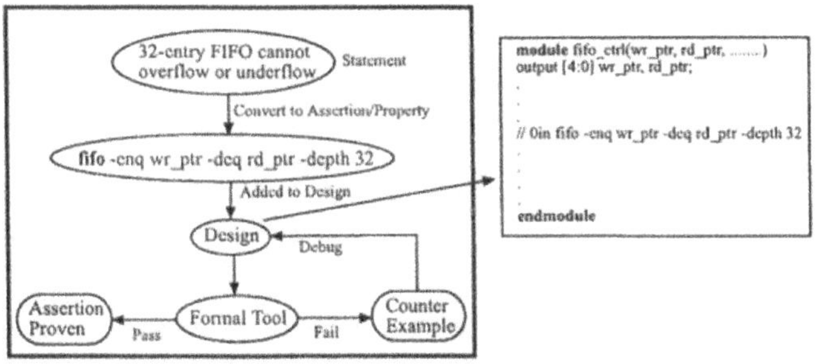

5. What is Mux Exclusivity and what are the most common techniques used to guarantee Mux Exclusivity in a design while in Scan mode?

Mux Exclusivity is guaranteeing the fact that not more than one bit in a mux select vector is active in a given cycle.

While scanning, selects and data for the pass gate muxes will be changing and there is a possibility that multiple mux selects could be active for few scan-in vectors unless ATPG tool (tool providing the test vectors) is burdened with the task of determining a contention free state for the mux selects. If the mux selects are not mutually exclusive then multiple selects will be simultaneously active causing contention on the output node thereby damaging the part. In a debug mode it is possible that this contentious state could persist for seconds or minutes or perhaps hours in the odd case. So it becomes necessary to guarantee mux exclusivity for a pass gate mux in the scan mode. Table below shows the most common techniques used to guarantee mux exclusivity while in scan mode. One advantage of technique 2 over technique 1 is that technique 2 results in better toggle coverage.

Table 48: Techniques Used to guarantee Mux Exclusivity

1	Adding mux protection logic after the flops which drive the mux selects. This can be achieved by gating the selects coming out of the flops with mux protection signal (active while in scan mode) as shown below. The mux protection signal which is active high in scan mode guarantees that the output of the mux protection logic is one hot in scan mode thereby guaranteeing mux exclusivity. This can be seen from figure below.
	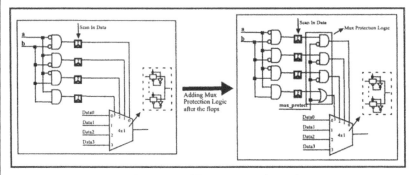
	Figure 153: Adding Mux Protection Logic
2	Moving select decode logic after the flops guarantees that the pass gate mux selects are mutually exclusive in scan mode. This can be seen from figure below.
	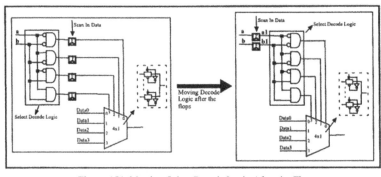
	Figure 154: Moving Select Decode Logic After the Flops

6. What is SAT (Stand Alone Testing) environment and what are the advantages of having SAT environment for Verification?

SAT is a verification environment where a Block is tested for functionality and coverage in a stand alone fashion. In some cases multiple Blocks instead of one could be tested. Figure below shows a typical SAT environment for a Block.

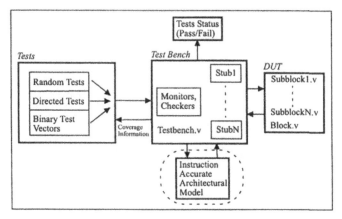

Figure 155: SAT Environment for a Block

Some of the advantages of having SAT environment are tabulated below.

Table 49: Advantages of SAT Environment

1	In early phases of the design where we don't have a full chip model, SAT provides a means for debugging.
2	By testing in SAT and releasing the code to full chip ensures good quality code to be released to full chip. This is important because debugging in full chip environment is much more cumbersome than debugging in SAT environment.
3	It is easy to setup test cases in SAT than in full chip environment (i.e better controlability and observability).
4	It provides an environment for independent work when the full chip model is broken.

7. What is Code Coverage and what are the various Code Coverage metrics used for Verification completeness?

Code coverage provides information on how thoroughly a design has been exercised during simulation. They are used to evaluate the effectiveness of random and directed tests and to guide the generation of new tests. Most common code coverage metrics used for verification completeness are tabulated below.

Table 50: Code Coverage Metrics used for Verification Completeness

Type of Coverage	Description

Block/ Statement Coverage	Block/Statement Coverage measures the percentage (i.e number of lines) of code that has been exercised by the test suite.
	
	Verilog code below has 9 statements (lines) that needs to be covered in order to have a 100% Block/Statement Coverage. If during simulation the test suite covers 7 (say lines 1, 2, 3, 4, 5, 7 and 8) out of 9 lines then we have a 77% Block/Statement Coverage for the code.
	
	<pre>wire [1:0] count_so, count_s1;
	assign r_stall_v1 = sq_full | lq_full;
	assign r_stall_v2 = cq_full | iq_full;
	always @(r_stall_v1 or r_stall_v2 or
	count_s0 or count_s1 or active_strand)
	  begin
	    if(r_stall_v1 | r_stall_v2)
	      switch_strand = 1'b1;
	    else
	      switch_strand = 1'b0;
	
	    if(active_strand)
	      count_as = count_s0;
	    else
	      count_as = count_s1;
	  end</pre>
	↓ Block/Statement Coverage
	
	|
		wire [1:0] count_so, count_s1;

line1	assign r_stall_v1 = sq_full | lq_full;
line2	assign r_stall_v2 = cq_full | iq_full;
line3	always @(r_stall_v1 or r_stall_v2 or count_s0 or count_s1 or active_strand)
	begin
line4	if(r_stall_v1 | r_stall_v2)
line5	switch_strand = 1'b1;
	else
line6	switch_strand = 1'b0;
line7	if(active_strand)
line8	count_as = count_s0;
	else
line9	count_as = count_s1;
	end	

Table 50: Code Coverage Metrics used for Verification Completeness

Type of Coverage	Description
Path Coverage	Path Coverage measures all possible ways you can execute a sequence of statements within a code. The paths through an **initial** or **always** verilog statement corresponds to the different sequence of statements that execute when initiating the **initial** or **always** statement. The **if, case/casex/casez**, and **disable** statements are the only verilog statements that can cause multiple paths to appear in a code. A sequence of statements contributes only one path when there are no **if, case/casex/casez**, and **disable** statements
	An **if** or **if....else** statement generates two paths (i.e one for the True condition and one for the False condition).
	A **case/casex/casez** statement has 'n' paths for 'n' case items.
	A **disable** statement placed within the block it names provides an early exit and defines two paths (i.e a normal exit and an early exit) through that block.
	Verilog code below has 4 paths that needs to be covered in order to have a 100% Path Coverage. If during simulation the test suite covers 2 out of 4 paths then we have a 50% Path Coverage for the code.

```
wire [1:0] count_so, count_s1;

assign r_stall_v1 = sq_full | lq_full;
assign r_stall_v2 = cq_full | iq_full;

always @(r_stall_v1 or r_stall_v2 or
count_s0 or count_s1 or active_strand)
  begin
    if(r_stall_v1 | r_stall_v2)
      switch_strand = 1'b1;
    else
      switch_strand = 1'b0;

    if(active_strand)
      count_as = count_s0;
    else
      count_as = count_s1;
  end
```

Path Coverage

(r_stall_v1 | r_stall_v2)

True   False

(active_strand)

4 Paths

Path1	False	False
Path2	False	True
Path3	True	False
Path4	True	True

True   False

Table 50: Code Coverage Metrics used for Verification Completeness

Type of Coverage	Description

Expression Coverage	Expression coverage ensures that the statements with expressions have been exercised to the fullest. Verilog expressions that get covered here are bitwise, reduction, logical, relational and event.
	
	Verilog code below has 14 coverage points for the expressions that need to be covered in order to have a 100% Expression Coverage. If during simulation the test suite covers 7 (shown in bold below) out of 14 coverage points then we have a 50% Expression Coverage for the code.
	
	<pre>wire [1:0] count_so, count_s1;
	assign r_stall_v1 = sq_full | lq_full;
	assign r_stall_v2 = cq_full | iq_full;
	always @(r_stall_v1 or r_stall_v2 or
	count_s0 or count_s1 or active_strand)
	  begin
	    if(r_stall_v1 | r_stall_v2)
	     switch_strand = 1'b1;
	    else
	      switch_strand = 1'b0;
	
	    if(active_strand)
	      count_as = count_s0;
	    else
	      count_as = count_s1;
	  end</pre>
	Expression Coverage
	
	|	

sq_full | lq_full **(0 | 0), (x | 1)**, (1 | x)	3
cq_full | iq_full **(0 | 0), (x | 1)**, (1 | x)	3
@(**r_stall_v1** or r_stall_v2 or **count_s0** or count_s1 or active_strand)	5
r_stall_v1 | r_stall_v2 **(0 | 0), (x | 1), (1 | x)**	3
	14	

Table 50: Code Coverage Metrics used for Verification Completeness

Type of Coverage	Description

Toggle Coverage	Toggle Coverage measures the ratio of the number of signals that experienced 1 to 0 and 0 to 1 transitions during simulation to the total number of effective signals. The number of effective signals here is adjusted to include only those that can be toggled.
	
	Verilog code below has 14 effective signals that needs to toggle (i.e from 1 to 0 and 0 to 1) in order to have a 100% Toggle Coverage. If during simulation the test suite results in toggling 5 out of 14 effective signals then we have a 35% Toggle Coverage for the code.
	
	<pre>wire [1:0] count_so, count_s1;
	
	assign r_stall_v1 = sq_full | lq_full;
	assign r_stall_v2 = cq_full | iq_full;
	
	always @(r_stall_v1 or r_stall_v2 or
	count_s0 or count_s1 or active_strand)
	  begin
	    if(r_stall_v1 | r_stall_v2)
	     switch_strand = 1'b1;
	    else
	     switch_strand = 1'b0;
	
	    if(active_strand)
	     count_as = count_s0;
	    else
	     count_as = count_s1;
	  end</pre>
	
	Toggle Coverage ⬇
	
	<pre>wire [1:0] count_so, count_s1;
	
	assign r_stall_v1 = sq_full | lq_full; [1] [2] [3]
	assign r_stall_v2 = cq_full | iq_full; [4] [5] [6]
	
	always @(r_stall_v1 or r_stall_v2 or
	count_s0 or count_s1 or active_strand)
	  begin
	    if(r_stall_v1 | r_stall_v2)
	     switch_strand = 1'b1; [7]
	    else
	     switch_strand = 1'b0;
	
	    if(active_strand) [8]
	     count_as = count_s0; [13,14] [9,10]
	    else
	     count_as = count_s1; [11,12]
	  end</pre>

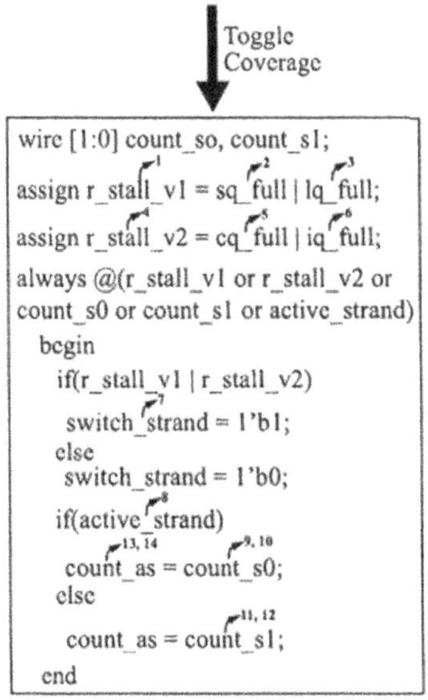

Table 50: Code Coverage Metrics used for Verification Completeness

Type of Coverage	Description
FSM Coverage	FSM's (Finite State Machine) are a special class of sequential logic. It consists of a combinational block that computes the next state for the next cycle and the output values for the current clock cycle and memory elements that preserve the present state of the machine.The next-state computation typically depends on the machine's present-state and input values. Output can be either Mealy or Moore. A Mealy output is a function of both present-state and inputs, while a Moore output is a function of only the present state of the machine. Typically FSM coverage involves State coverage, Arc coverage and Sequence coverage. State coverage measures the number of states within an FSM exercised by the test suite, Arc coverage measures the number of valid arcs exercised by the test suite and Sequence coverage measures the valid sequences exercised by the test suite. Figure below shows a simple Strand/Thread switch FSM. In order to have a 100% State coverage for the FSM the test suite should exercise all 3 states (i.e **S0** (Strand0), **S1** (Strand1) and **S2** (Strand2)). If the test suite exercises 2 out of 3 states then we have a 66% State coverage. In order to have a 100% Arc coverage for the FSM the test suite should exercise all 6 arcs shown below. If the test suite covers 2 out of 6 arcs then we have a 33% Arc coverage. In order to have a 100% Sequence coverage for the FSM the test suite should exercise all 15 valid sequences i.e **S0, S0 -> S1, S0 -> S2, S0 -> S1 -> S2, S0 -> S2 -> S1, S1, S1 -> S0, S1 -> S2, S1 -> S0 -> S2, S1 -> S2 -> S0, S2, S2 -> S0, S2 -> S1, S2 -> S0 -> S1 and S2 -> S1 -> S0.** If the test suite results in exercising 4 out of 15 sequences then we have a 26% Sequence coverage. Figure 156: Strand/Thread Switch FSM

8. Describe LBIST, MBIST and ATPG?

LBIST

LBIST (Logic Built-In Self Test) tests on chip logic at speed. Here thousands of tests can be executed at speed without consuming any tester memory. Typically LBIST is activated on every Power-On-Reset (POR). A typical LBIST architecture is shown in the figure below

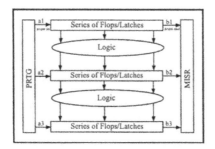

Figure 157: LBIST Architecture

In the above PRTG (Pseudo Random Test Generator) is responsible for generating test patterns for logic which needs to be tested using LBIST. In many cases PRTG is a maximal length LFSR (Linear Feedback Shift Register). MISR (Multiple Input Signature Analyzer Register) is responsible for collecting and compressing the test responses. Sequence of operations in LBIST are: fill in the scan chains with data from PRTG; transition to functional mode and advance your clock; transition to scan mode and scan out the results of the test into MISR for signature generation while simultaneously filling in the chains with new data from PRTG. When the entire testing is complete the MISR will hold a signature that can be compared against the expected signature for a fault free logic. A very simple controller can be used to initiate LBIST. The controller itself can be initiated using a special JTAG instruction, through software or through dedicated pins on chip.

Eventhough LBIST provides good fault coverage (ratio of the number of faults detected to the total number of possible faults), it doesn't provide enough fault coverage to replace manufacturing tests driven by ATPG scan patterns.

MBIST

MBIST (Memory Built-In Self Test) tests on chip memories at speed. Here tests can be executed at speed without consuming any tester memory. Typically MBIST is activated on every POR. A simple MBIST architecture is shown in the figure below.

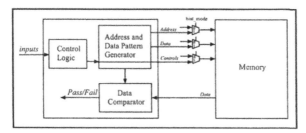

Figure 158: MBIST Architecture

In the above Address and Data Pattern Generator is responsible for providing the address and data to memory blocks which needs to be tested. Control logic provides the necessary Read/Write and other control information. Data comparator compares the read data from memory with the expected data and provides Pass/Fail information based on whether there was a mismatch or not. Few BIST controllers support capturing the failing data/address which can be scaned out or read through software. Typical algorithms used for address and data generation are 6N, March, March C etc. March C algorithm is shown below

$$\{ \downarrow WO; \downarrow R0W1; \downarrow R1W0; \downarrow R0W1; \downarrow R1W0; \downarrow R0 \}$$
marching element

Here the test is composed of six elements: the first element traverses the memory address space in the ascending order writing the '0' data pattern, the second element traverses the address space in the ascending order reading the '0' data pattern and writing the ' 1' data pattern. The sequence of read, write address change is as follows: at the first address location the BIST controller will read '0' data pattern and at the same location writes the '1' data pattern, the engine will then increment the address to the next location and repeat the read/write process. Testing continues in a similar fashion for the remainder of the address space in the descending order. Here a '0' data pattern need not necessarily be all '0's and a '1' data pattern need not necessarily be all 1's. Some of the data patterns used are (8'h66, 8'h99), (8'h33, 8'hCC), (8'h55, 8'hAA) etc. In some cases BIST controllers provide an option where user can scan in his data which he wants to be used as a data pattern.

Here the controller can be initiated using a special JTAG instruction, through software or through dedicated pins on chip.

ATPG

ATPG (Automatic Test Pattern Generation) is the application of a set of algorithmic techniques to generate a set of test patterns that detects faulty behavior of a circuit after its fabrication (i.e manufacturing defects). Figure below shows a ATPG pattern applied to a combinational logic that detects stuck-at fault in the logic.

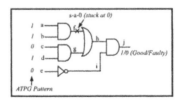

Figure 159: ATPG Pattern Detecting Stuck-At Fault

In a scan based architecture, testing using ATPG patterns typically involves the following steps: scan in the ATPG pattern into the sequential elements (i.e flops/latches) while in scan mode; revert to functional mode and advance your clock; revert back to scan mode and scan out the data while concurrently scaning in a new ATPG pattern. If the scanned out data does not match with the expected data pattern then we have a fault in the circuit.

9. What are Spare Gates and what are the advantages of having Spare Gates in a Design?

Spare gates are additional non functional gates (i.e gates not involved in functionality of a block) sprinkled in a block.

Table below lists the advantages of having spare gates in a design.

Table 51: Advantages of having Spare Gates in a Design

	Advantages
1	Late bugs before tapeout can be fixed in many cases using Spare Gates. This prevents us from respinning (i.e going through the entire backend flow (i.e synthesis, floorplanning, place & route etc.)) the block all over again.
2	They can be used to fix bugs found during post silicon debug effort. This allows us in preparing metal only masks instead of preparing all layer masks thereby saving lots of $$$$.
3	They can be used to do FIB (focussed ion beam (it uses an ion beam to cut and expose various metal lines on a functional chip and to deposit platinum thereby reconnecting the gates into a new logic structure)) for the bug in lab before committing to a second tapeout with fix for the bug in place. FIBing is one way of guaranteeing that the fix for the bug is going to work or not.

A rule of thumb is to have 3% of the total gate count as spare gate count within a block. A typical distribution for the spare gates is 20% Flops, 30% inverters and 50% complex and non-complex gates

This page intentionally left blank

5 Tools

1. What are the most common Software Verilog Simulation tools used in the industry?

Software Verilog simulation tools are used to compile and simulate designs described in Verilog. Table below shows the most common software Verilog simulation tools used in the industry.

Table 52: Verilog Simulation Tools

Event-driven Tool	Cycle-based Tool	Company
NC-Verilog, Verilog-XL	SpeedSim	Cadence
VCS	Polaris	Synopsys
ModelSim	X	Mentor Graphics

2. What are the most common Software VHDL Simulation tools used in the industry?

Software VHDL simulation tools are used to compile and simulate designs described in VHDL. Table below shows the most common software VHDL simulation tools used in the industry.

Table 53: VHDL Simulation Tools

Event-driven Tool	Cycle-based Tool	Company
NC-VHDL	X	Cadence
Scirocco	Scirocco	Synopsys
ModelSim	X	Mentor Graphics

3. What are the most common Linting tools used in the industry?

Linting tools are used to make sure that designs coded in Verilog and VHDL are coded properly. Most of the Processor companies have there own inhouse linting tools. Table below shows the most common commercially available Linting tools used in the industry.

Table 54: Linting Tools

Tool	Company
VN-Check	TransEDA
Verity-Check	Veritable
0-In Checklist	0-In

4. What are the most common Testbench Automation tools used in the industry?

Testbench Automation tools are used to automate a testbench environment for a design to be tested. Table below shows the most common Testbench Automation tools used in the industry.

Table 55: Testbench Automation Tools

Tool	Company
Vera	Synopsys
Specman Elite	Verisity

5. What are the most common languages used to develop a Testbench?

The most common languages used to develop a Testbench are OpenVera, e, Verilog, VHDL and C.

6. What are the most common Debugging tools used in the industry?

Debugging tools help in debugging a design. Table below shows the most common Debugging tools used in the industry.

Table 56: Debugging Tools

Tool	Company
Signalscan	Cadence

Table 56: Debugging Tools

Tool	Company
Debussy	Novas
Undertow	Veritools

7. What are the most common Formal Functional Verification tools used in the industry?

Formal Functional Verification tools help find bugs in a design. Table below shows the most common Formal Functional Verification tools used in the industry.

Table 57: Formal Functional Verification Tools

Tool	Company
0-in Search	0-In
BlackTie UDC	Cadence
Magellan	Synopsys
VN-Property	TransEDA

8. What are the most common Formal Equivalence Checking tools used in the industry?

Formal Equivalence Checking tools verify equivalency between two designs (i.e RTL against Gate level netlist etc.). Table below shows the most common Formal Equivalence Checking tools used in the industry.

Table 58: Formal Equivalence Checking Tools

Tool	Company
Conformal LEC, FormalCheck	Cadence
Formality, Design Verifyer	Synopsys
FormalPro	Mentor Graphics
ESP-CV, Innologic	

9. What are the most common Code Coverage tools used in the industry?

Code coverage tools are used to provide coverage information for a design being tested. Table below shows the most common Code Coverage tools used in the industry.

Table 59: Code Coverage Tools

Tool	Company
SureCove	Verisity
VN-Cover	TransEDA
VeriCover	Veritools
HDLScore	Innoveda
VCS	Synopsys
NC-Verilog, NC-VHDL	Cadence
Covermeter	Synopsys

10. What are the most common Hardware Accelerator tools used in the industry?

They are custom hardware dedicated to simulate through high speed RAM and proprietary ASIC designs architected for parallel, pipelined operation. The main difference between acceleration and emulation is what drives the design stimulus, which ultimately impacts the run-time performance. With acceleration, stimulus is provided by a software testbench whereas with emulation stimulus comes from live electrical connections. With acceleration if stimulus is residing in a workstation then the accelerator need to synchronize at every clock cycle. Table below shows the most common Hardware Accelerator tools used in the industry.

Table 60: Hardware Accelerator Tools

Tool	Company
Xcite, XoC, Xtreme	Axis Systems
Hammer	Tharas Systems
Palladium, CoBALT Ultra	Cadence
ARES	Mentor Graphics
HES	Alatek

11. What are the most common Hardware Emulation tools used in the industry?

Hardware Emulation tools are used to emulate a design in hardware using ASIC's and FPGA's. Table below shows the most common Hardware Emulation tools used in the industry.

Table 61: Hardware Emulation Tools

Tool	Company
XoC, Xtreme	Axis Systems
Palladium, CoBALT Ultra	Cadence
VStation, CelaroPRO	Mentor Graphics
COMULATOR	Alatek

12. What are the most common Logic Synthesis tools used in the industry?

Logic Synthesis tools are used to synthesize a design written in Verilog or VHDL. Table below shows the most common Logic Synthesis tools used in the industry.

Table 62: Logic Synthesis Tools

Tool	Company
DesignCompiler	Synopsys
Blast Create	Magma
BuildGates	Cadence
Synplify ASIC	Synplicity

13. What are the most common FPGA Synthesis tools used in the industry?

FPGA Synthesis tools are used to synthesize a design for it to be mapped to FPGA's. Table below shows the most common FPGA Synthesis tools used in the industry.

Table 63: FPGA Synthesis Tools

Tool	Company
Synplify Pro	Synplicity
PrecisionRTL, LeonardoSpectrum	Mentor Graphics

Table 63: FPGA Synthesis Tools

Tool	Company
FPGA Compiler II	Synopsys
XST	Xilinx

14. What are the most common Static Timing Analysis tools used in the industry?

Timing analysis is required to verify the timing performance of a design by ensuring that the setup and hold times of flip-flops/latches are met and the critical paths in the design meet the required timing spec. Static timing analysis tools are used for this purpose. They analyze all the paths in the design to see if they meet the required timing spec. Few advantages of Static timing analysis over Dynamic timing analysis is that they are a magnitude faster than Dynamic timing analysis tools and they don't require vectors for simulation. One of the main disadvantage of these tools is that they result in reporting false paths (these are the paths which are reported as critical but in reality they are non-critical as these paths are never exercised during normal operation of the circuit/logic).Table below shows the most common Static Timing Analysis tools used in the industry.

Table 64: Static Timing Analysis Tools

Tool	Company
PrimeTime, PathMill	Synopsys
Pearl, PKS	Cadence
Blast Logic	Magma
SST Velocity	Mentor Graphics
ShowTime	Sequence Design
Dolphin	Monterey Design Systems
DynaCore	Circuit Semantics

15. What are the most common Dynamic Timing Analysis tools used in the industry?

Dynamic Timing Analysis tools like Static tools are used to verify timing performance of a design. One of the key differences between Dynamic and Static tools is that Dynamic Timing Analysis tools require vectors for simulation and they don't report any false paths. Table below shows the most common Dynamic Timing Analysis tools used in the industry.

Table 65: Dynamic Timing Analysis Tools

Tool	Company
Spectre, PSpice	Cadence
Mach TA	Mentor Graphics
NanoSim, Star-SimXT, TimeMill, HSPICE	Synopsys
SmartSpice	Silvaco

16. What are the most common Power Analysis tools used in the industry?

Power Analysis tools are used to analyze power consumed by a design. Table below shows the most common Power Analysis tools used in the industry.

Table 66: Power Analysis Tools

Gate/Circuit Power Analysis Tool	RTL Power Analysis Tool	Company
PowerMill, PrimePower	Power Compiler	Synopsys
Blast Rail	X	Magma
X	PowerTheater Designer	Sequence Design

17. What are the most common Schematic Capture tools used in the industry?

Schematic Capture tools are used to capture schematics for a design. Table below shows the most common Schematic Capture tools used in the industry.

Table 67: Schematic Capture Tools

Tool	Company
Design Architect-IC	Mentor Graphics
Virtuoso Composer, OrCAD Capture	Cadence
CosmosSE	Synopsys
Scholar	Silvaco

18. What are the most common Place and Route tools used in the industry?

Place and Route tools are used to place and route standard cells in a design. Table below shows the most common Place and Route tools used in the industry.

Table 68: Place and Route Tools

Tool	Company
Silicon Ensemble	Cadence
Apollo, Astro	Synopsys
AutoCells	Mentor Graphics
Blast Fusion	Magma
Dolphin	Monterey Design Systems

19. What are the most common Floorplanning tools used in the industry?

Floorplanning tools help in floorplanning a design. Table below shows the most common Floorplanning tools used in the industry.

Table 69: Floorplanning Tools

Tool	Company
Chip Architect	Synopsys
SoC Encounter	Cadence
IC Wizard	Monterey Design Systems
Blast Plan	Magma
IC Station	Mentor Graphics

20. What are the most common Layout Editor tools used in the industry?

Layout Editor tools help in drawing mask layers for a design. Table below shows the most Layout Editor tools used in the industry.

Table 70: Layout Editor Tools

Tool	Company
IC Station	Mentor Graphics
Virtuoso-XL Layout Editor	Cadence
CosmosLE	Synopsys
Expert	Silvaco

21. What are the most common Chip Level Routing tools used in the industry?

Chip Level Routing tools are used to route metal layers at the Chip Level. Table below shows the most common Chip Level Routing tools used in the industry.

Table 71: Chip Level Routing Tools

Tool	Company
FlexRoute	Synopsys
Virtuoso Chip Assembly Router, NanoRoute Ultra	Cadence
IC Station	Mentor Graphics

22. What are the most common Physical Verification tools used in the industry?

Physical Verification tools are used to verify LVS (Layout versus Schematic), DRC (Design Rule Checking) and ERC (Electrical Rule Checking) for a design. Table below shows the most common Physical Verification tools used in the industry.

Table 72: Physical Verification Tools

Tool	Company
Dracula, Diva, Assura DRC, Assura LVS	Cadence
ICVerify, Calibre	Mentor Graphics
Hercules	Synopsys
Guardian	Silvaco

23. What are the most common Parasitic Extraction tools used in the industry?

Parasitic Extraction tools are used to extract parasitics (i.e Resistance and Capacitance values) for a design. Table below shows the most common Parasitic Extraction tools used in the industry.

Table 73: Parasitic Extraction Tools

Tool	Company
Fire & Ice QXC, Assura RCX	Cadence
Star-RCXT	Synopsys
Calibre xRC	Mentor Graphics
Blast Fusion	Magma
Clever	Silvaco
Columbus-AMS	Sequence Design

24. What are the most common Noise Analysis tools used in the industry?

Noise Analysis tools are used to analyze noise in a design. Table below shows the most common Noise Analysis tools used in the industry.

Table 74: Noise Analysis Tools

Tool	Company
PrimeTime SI	Synopsys
Signal Storm	Cadence
Blast Noise	Magma
CoolTime	Sequence Design

25. What are the most common EM-IR Analysis tools used in the industry?

EM-IR tools verify for IR drop and EM violations on power and ground networks. Table below shows the most common EM-IR tools used in the industry.

Table 75: EM-IR Analysis Tools

Tool	Company
Thunder, Hail	Cadence
RailMill	Synopsys
Blast Rail	Magma
Physical Studio, CoolTime	Sequence Design

26. What are the most common ATPG (Automatic Test Pattern Generator) tools used in the industry?

ATPG tools are used to provide test vectors for a design. Table below shows the most common ATPG tools used in the industry.

Table 76: ATPG Tools

Tool	Company
FastScan	Mentor Graphics
TetraMAX	Synopsys
TurboScan	Syntest

27. What are the most common Boundary Scan tools used in the industry?

Boundary Scan tools are used to verify connectivity at the module level. Table below shows the most common Boundary Scan tools used in the industry.

Table 77: Boundary Scan Tools

Tool	Company
BSDArchitect	Mentor Graphics
BSD Compiler	Synopsys
TurboBSD	Syntest
Eclipse	Intellitech

Table 77: Boundary Scan Tools

Tool	Company
ScanPlus	Corelis

28. What is the difference between a cycle-based simulator and a event-driven simulator?

Cycle-based simulation is generally faster than event-driven simulation. It is used most effectively early in the design cycle where timing verification is not yet critical. Together with equivalence checking and static timing analysis, cycle-base simulation can provide a complete sign-off methodology without the need for gate-level simulation. Event-driven simulation continues to dominate the sign-off verification marketplace and will continue to do so until more trust is built by using cycle-based simulation, equivalence checking and static timing analysis as a sign-off methodology. Trust will be a function of the relative number of successful chips built using this newer methodology. While it will take several years before this trust is built, event-driven simulation will have a central role. It is expected that the use of event-driven simulation will continue to increase even as the shift to the newer methodology is made. Even in the future, after sign-off methodologies are radically changed, event-driven simulation will still be useful in many circumstances i.e asynchronous blocks where timing and function are closely intertwined, for example.

6 Verilog

1. What are the various Operators in Verilog?

Table below lists the various Operators in Verilog.

Table 78: Operators in Verilog

Op	Description	Usage		Op	Description	Usage
Arithmetic Operators (It takes two arguments which are one or more bits long and returns a result which is one or more bits long)				*Logical Operators* (The result is a 1-bit True/False)		
+	Addition	(4'b0101) + (4'b0101) = 4'b1010		&&	Logical AND	(4'b1001) && (4'b0001) = 1'b1 (4'b1001) && (4'b0000) = 1'b0
-	Substraction	(4'b0101) - (4'b0101) = 4'b0000		||	Logical OR	(4'b1001) || (4'b0001) = 1'b1 (4'b1001) || (4'b0000) = 1'b1
*	Multiplication	(4'b0001) * (4'b0011) = 4'b0011		!	Logical NOT	!(4'b1001) = 1'b0 !(4'b0000) = 1'b1
/	Division	(4'b0100) / (4'b0010) = 4'b0010				
%	Modulus	(4'b0101) % (4'b0101) = 4'b1010				
Bitwise Operators (The result is a 1-bit True/False)				*Reduction Operators* (They operate on only one operand. They operate on one or more than one bit operand and return a single bit result)		
~	NOT	~(4'b0101) = 4'b1010		&	Reduction AND	&(4'b1001) = 1'b0 &(4'b1111) = 1'b1
&	AND	(4'b1001) & (4'b1011) = 4'b1001		~&	Reduction NAND	~&(4'b1001) = 1'b1 ~&(4'b1111) = 1'b0
|	OR	(4'b1001) | (4'b1011) = 4'b1011		|	Reduction OR	|(4'b1001) = 1'b1 |(4'b0000) = 1'b0
^	EX-OR	(4'b1001) ^ (4'b1011) = 4'b0010		~|	Reduction NOR	~|(4'b1001) = 1'b0 ~|(4'b0000) = 1'b1

Table 78: Operators in Verilog

Op	Description	Usage	Op	Description	Usage
~^	EX-NOR	(4'b1001) ~^ (4'b1011) = 4'b1101	^	Reduction EX-OR	^(4'b1001) = 1'b0 ^(4'b1000) = 1'b1
<<	Shift Left	(4'b1001) << (4'b0001) = 4'b0010	~^	Reduction EX-NOR	~^(4'b1001) = 1'b1 ~^(4'b1000) = 1'b0
>>	Shift Right	(4'b1001) >> (4'b0001) = 4'b0100			
	Equality Operators (The result is a 1-bit True/False)			*Identity Operators* (The result is a 1-bit True/False)	
==	Equivalent	(4'b1001) == (4'b0001) = 1'b0 (4'b1001) == (4'b1001) = 1'b1 (4'b1001) == (4'b000x) = 1'bx (4'b1001) == (4'b000z) = 1'bx	===	Identical	(4'b1001) === (4'b0001) = 1'b0 (4'b1z01) === (4'b1z01) = 1'b1 (4'b1001) === (4'b000x) = 1'b0 (4'b100x) === (4'b000x) = 1'b1
!=	Non-Equivalent	(4'b1001) != (4'b0001) = 1'b0 (4'b1001) != (4'b1001) = 1'b1 (4'b1001) != (4'b000x) = 1'bx (4'b1001) != (4'b000z) = 1'bx	!==	Non_Identical	(4'b1001) !== (4'b0001) = 1'b1 (4'b1z01) !== (4'b1z01) = 1'b0 (4'b1001) !== (4'b000x) = 1'b1 (4'b100x) !== (4'b000x) = 1'b0
	Other Operators (The result is a 1-bit True/False)			*Relational Operators* (The result is a 1-bit True/False)	
{}	Concatenate	{4'b1000, 4'b0001} = 8'b10000001	<	Less than	(4'b1001) < (4'b0001) = 1'b0 (4'b1001) < (4'b1101) = 1'b1
{{}}	Replicate	{2{4'b1000}} = 8'b10001000	>	Greater than	(4'b1001) > (4'b0001) = 1'b1 (4'b1001) > (4'b1101) = 1'b0
? :	Conditional	out = s ? a : b; out = a if s=1'b1 out = b if s=1'b0	<=	Less than or Equal to	(4'b1001) <= (4'b0001) = 1'b0 (4'b1001) <= (4'b1001) = 1'b1
			>=	Greater than or Equal to	(4'b1001) >= (4'b0001) = 1'b1 (4'b1001) >= (4'b1001) = 1'b1

2. What is the Operator Order of Precedence in Verilog?

Table below lists the Operator Order of Precedence in Verilog.

Table 79: Operator Order of Precedence

Operators	Description	Precedence
! ~	Negation	Highest Precedence
* / %	Arithmetic, Multiplication, Division	
+ -	Addition, Substraction	
<< >>	Shift Left, Shift Right	
< <= > >=	Relational	
== != === !==	Equality	
& ~&	Bitwise AND, Bitwise NAND	
^ ~^	Exclusive OR, Exclusive NOR	
| ~|	Bitwise OR, Bitwise NOR	
&&	Logical AND	
||	Logical OR	
?:	Ternary	Lowest Precedence

3. List some of the Format Specifiers in Verilog?

Table below lists some of the Format Specifiers in Verilog.

Table 80: Format Specifiers

Symbol	Description	Symbol	Description
%b	Binary with leading zeroes	%c	Character
%0b	Binary with no leading zeroes	%s	String
%d	Decimal	%t	Time format
%0d	Decimal with leading spaces truncated	%f	Real in decimal format
%h	Hexadecimal with leading zeroes	%e	Real in exponential format
%0h	Hexadecimal with no leading zeroes	\n	New line

4. What are System Tasks and Functions in Verilog?

There are certain buit-in commands in Verilog to provide support for system functions such as printing messages or reading and writing of files etc. These special commands are known as System Tasks and Functions and they always begin with "$" symbol. Users may define their own built-in tasks and functions using Verilog Programming Language Interface. Table below lists the most commonly used System Tasks and Functions.

Table 81: System Tasks and Functions

System Task	Description	Usage
$display	Prints the formatted message once when the statement is executed during simulation. A new line is automatically added at the end of its output.	$display("optional text with format specifier", signal1, signal2, ..); Example initial $display("Hello World"); Prints the string between the quotation marks with a new line added at the end of the string
$write	Prints the formatted message once when the statement is executed during simulation. No newline is added at the end of its output.	$write("optional text with format specifier", signal1, signal2, ..); Example initial $write("Hello World"); Prints the string between the quotation marks with no new line added at the end of the string
$strobe	It executes after all simulation events in the current time step have executed. Prints the formatted message once when executed. This task guarantees that the printed values for the signals/variables are the final values the signals/variables can have at that time step.	$strobe("optional text with format specifier", signal1, signal2, ..); Example initial $strobe("Current values of A, B and C are A=%b, B=%b, C=%b", A, B, C); Prints A, B and C and prints their value in binary format after all simulation events in the current time step have executed.

Table 81: System Tasks and Functions

System Task	Description	Usage
$monitor	Invokes a background process that continuously monitors the signals listed, and prints the formatted message whenever one of the signals changes. A newline is automatically added at the end of its output. Only one **$monitor** can be active at a time.	**$monitor**("optional text with format specifier", signal1, signal2, ..); Example initial **$monitor**("Current values of A, B and C are A=%b, B=%b, C=%b", A, B, C); Monitors A, B and C and prints their value in binary format whenever one of the signals(i.e A or B or C) changes its value during simulation.

$fopen	A function that opens a file to redirect the output of Verilog. It returns a 32-bit value called a multi-channel descriptor(mcd). Only one bit is set in a mcd. The standard output(also known as channel 0) has a mcd with the LSB(i.e bit 0) set. Each successive call to $fopen opens a new channel and returns a 32-bit mcd with bit 1 set, bit 2 set and so on upto bit 31 set.	mcd = **$fopen**("name_of_file");
		
		Example
		initial f1 = **$fopen**("error_file");
		
		Opens a file named 'error_file' and provides a file handler(i.e 'f1') by setting one of the bits in the 32-bit mcd to 1.

$fclose	A function that closes a file that was opened by $fopen.	**$fclose**(mcd); Example initial **$fclose**("f1"); Closes file corresponding to file handler 'f1' that was opened by $fopen.
$fdisplay	Writes to a file the formatted message once when the statement is executed during simulation. A new line is automatically added at the end of its output.	**$fdisplay**(mcd, "optional text with format specifier", signal1, signal2, ..); Example initial **$fdisplay**(f1 , "Current values of A, B and C are A=%b, B=%b, C=%b", A, B, C); Writes to a file referred by 'f1' the formatted text along with the binary values of A, B and C once the statement gets executed.

Table 81: System Tasks and Functions

System Task	Description	Usage
$timeformat	Controls the format used by the %t text format specifier.	**$timeformat**(*unit, precision, "suffix", field_width*); *unit* - base that time is to be displayed in(-9 = 1ns, -12= 1ps etc.) *precision* - specifies number of decimal points to be displayed *suffix* - string appended to the time("ns", "ps", "nanoseconds" etc.) *field_width* - minimum number of characters that will be displayed Example initial **$timeformat**(-12, 2, "ps",10); Displays values like 20.00ps, 40.00ps etc.
$time	Returns the current simulation time as a 64-bit integer.	**$time** Example initial $monitor("time = %d, A = %d", **$time**, A); Prints the current simulation time as a 64-bit decimal whenever $monitor gets executed.
$stime	Returns the lower 32-bits of simulation time as an integer.	**$stime** Example initial $monitor("time = %d, A = %d", **$stime**, A); Prints the lower 32-bits of the current simulation time as a 32-bit decimal whenever $monitor gets executed.

$realtime	Returns a real number representation of simulation time.	**$realtime**
		
		Example
		initial
		  begin
		    $timeformat(-12, 2, "ps", 10);
		    $monitor("%t, A = %d", **$realtime**, A);
		  end
		
		Prints a real number representation of simulation time.

Table 81: System Tasks and Functions

System Task	Description	Usage
$random	Returns a random 32-bit signed random number. It can be used with a seed to ensure the same random number sequence each time the test is run.	**$random;** **$random**(seed); Example initial **$random;** initial **$random**(2356);
$finish	Finishes a simulation and exits the simulation process.	**$finish;** Example initial **$finish;**
$stop	Halts a simulation and enters an interactive debug mode.	**$stop;** Example initial **$stop;**
$readmemb	Opens a file for reading, and loads the contents into a register memory array. The file must be an ASCII file with values represented in binary. The start and end address values are optional.	**$readmemb**("file_name", memory_name); **$readmemb**("file_name", memory_name, start_addr); **$readmemb**("file_name", memory_name, start_addr, finish_addr); Example initial **$readmemb**("initialize", dcache); The file 'initialize' contains the initialization data for memory array 'dcache'. Addresses are specified in the file with @address. They are specified as hexadecimal numbers. Data is separated by white spaces and may contain 'X' or 'Z'. Uninitialized locations default to 'X'. A sample data for 'initialize' is @004 11111111 00000000 @006 1111zzzz zzzz0000

Table 81: System Tasks and Functions

System Task	Description	Usage
$readmemh	Opens a file for reading, and loads the contents into a register memory array. The file must be an ASCII file with values represented in hex. The start and end address values are optional.	**$readmemh**("file_name", memory_name); **$readmemh**("file_name", memory_name, start_addr); **$readmemh**("file_name", memory_name, start_addr, finish_addr); Example initial **$readmemh**("initialize", dcache); The file 'initialize' contains the initialization data for memory array 'dcache'. Addresses are specified in the file with @address. They are specified as hexadecimal numbers. Data is separated by white spaces and may contain 'X' or 'Z'. Uninitialized locations default to 'X'. A sample data for 'initialize' is @004 AA CC @006 D1 23
$dumpvars	A system task which is used to dump value-changes of variables into a file which gets used by waveform viewer to display the results in the form of a waveform display.	**$dumpvars**(levels, list_of_modules); Example initial **$dumpvars**(2, top); Dumps all variables upto 2 levels of hierarchy in module instance 'top'.
$dumpfile	A system task which allows you to provide a file name where the value-changes of variables needs to be dumped to.	**$dumpfile**("file_name"); Example initial **$dumpfile**("signals.dump"); Dumps the value-changes of variables into a file called 'signals.dump'.
$dumpon	A system task used to control the start of dumping process.	**$dumpon**; Example initial #100 **$dumpon**; Start dump process after 100 timeunits.

Table 81: System Tasks and Functions

System Task	Description	Usage
$dumpoff	A system task used to control the stop of dumping process.	**$dumpoff;** Example initial #500 **$dumpoff;** Stop dump process after 500 timeunits.

5. What are Compiler Directives in Verilog?

Compiler Directives direct the pre-processor part of Verilog Parser or Simulator. Some of the processing involves substitution of strings, conditional inclusion and exclusion of code etc. They all precede with the character ('). They ar not bound by modules or files. When a Simulator encounters a compiler directive, the directive remains in effect until another compiler directive either modifies it or turns it off. Table below lists the most commonly used Compiler Directives.

Table 82: Compiler Directives

Compiler directive	Description	Usage
'include	File inclusion. The contents of another Verilog source file is inserted where the 'include directive appears.	'include "file name" Example *module* idu() ... 'include define.h ... *endmodule* Here the contents of 'define.h' are included within the 'idu' module.
'define	Allows a text string to be defined as a macro name.	'define macro_name text_string Example 'define *gate_regression* 1 Allows '*gate_regression*' to be substituted by 1 where ever it gets used.

Table 82: Compiler Directives

Compiler directive	Description	Usage
'ifdef-'else-'endif	Conditional compilation. Allows Verilog source code to be optionally included based on whether or not the macro_name has been defined using 'define or an invocation option.	'ifdef-'else-'endif Example 'define *gate_regression* 1 'ifdef *gate_regression* (gate_source_code) 'else (rtl_source_code) 'endif Here since 'gate_regression' is 1 we end up compiling 'gate_source_code' instead of 'rtl_source_code'.
'timescale	Specifies the time units and precision for delays, 'time unit' is the amount of time a delay of 1 represents. The time unit must be 1, 10 or 100. 'base' is the time base for each unit, ranging from seconds to femtoseconds, and must be s, ms, us, ps or fs. 'precision' represents how many decimal points of precision to use relative to the time units.	'timescale time_unit base/precision base Example 'timescale 1ns/1ns initial #100 assign a = b & c; Here 'a' gets evaluated after 100 time units(i.e 100ns).
'resetall	Resets all compiler directives back to its original default values.	'resetall
'defaultnettype	Changes the net data type to be used for implicit net declarations in a design.	'defaultnettype

6. What are Procedural blocks in Verilog?

Procedural blocks in Verilog are used to model both combinatorial and sequential logic. They are also used in building a test bench environment for a design. There are two types of Procedural blocks which are *'initial'* and *'always'*.

'initial' procedural block starts at time 0 and executes exactly once during a simulation and then does not execute again. If there are multiple *'initial'* blocks then each block starts to execute concurrently at time 0. Each *'initial'* block finishes execution independently of other blocks. The time at which they finish depends on the code in the *'initial'* block. It is possible to have a *'initial'* block that never finishes. An example of *'initial'* block is shown below,

```
parameter period = 500;
initial
  begin
   clock = 0;
   forever #(period/2) clock = ~clock;
  end
```

'always' procedural block is similar in behavior to 'initial' block except that it exe-cutes the statements repeatedly(i.e once the statements finish execution they start exe-cuting all over again). An 'always' block is like an 'initial' block with an infinite loop. An example of 'always' block is shown below,

```
always@(posedge clock)
  begin
   read_f <= read;
   write_f <= write;
   data_in_f <= data_in;
   index_f <= address_in;
  end
```

'initial' and 'always' procedural blocks cannot be nested.

7. What are the Timing Control Statements in Verilog?

Table below lists the Timing Control Statements supported in Verilog.

Table 83: Timing Controls in Verilog

Timing Controls	Description
#	Syntax #delay It delays execution for a specific amount of time, 'delay' may be a number, a variable or an expression. Example of a procedural block using '#' is shown below.

Table 83: Timing Controls in Verilog

Timing Controls	Description

	``
	always
	  begin
	    a = #2 (c & d);
	    #3 e = (f | g);
	  end
	``
	
	Here '(c & d)' gets evaluated at time 0 but gets assigned to 'a' after 2 time units whereas (f | g) gets evaluated after 3 timeunits and gets assigned to 'e' immediately.

@	<u>Syntax</u>
	@(*edge signal* or *signal* or ....)
	
	It delays execution until there is a transition on any one of the signals in the sensitivity list. *'edge'* may be either a posedge or negedge. If no edge is specified then any logic transition is used. Signals here may be scalar or vector, and any data type. Example of a procedural block using '@' is shown below.
	
	``
	always@(b or c)
	  a = b & c;
	``
	
	Here the statement within the always block gets evaluated when ever there is a transition on 'b' or 'c'.

wait	<u>Syntax</u>
	wait(*condition*)
	
	It delays execution until the condition evaluates as true. Example of a procedural block using 'wait' is shown below.
	
	``
	Initial
	  begin
	    wait(a == 2)
	      c = f & g;
	  end
	``
	
	Here we wait for 'a' to be equal to 2 before evaluating 'e'.

8. What are the various Programming Statements used in Verilog?

Table below lists the various Programming Statements used in Verilog.

Table 84: Programming Statements in Verilog

Programming Statements	Usage
if	<u>Syntax</u> If(*condition*) *statement* or *statement_group* It executes the *statement* or *statement_group* if the condition evaluates as true. If we need more than one statement (i.e *statement_group*) then we need to use *begin-end* or a *fork-join* block. The *condition* here can be an expression or a single value. If the *condition* evaluates to '0' or unknown then the *condition* is considered false, and if the *condition* evaluates to '1' or more then the *condition* is considered true.

if-else	<u>Syntax</u>
	If(*condition*)
	    statement or *statement_group*
	else
	    statement or *statement_group*
	
	It executes the first *statement* or *statement_group* if the condition evaluates as true and executes the second *statement* or *statement_group* if the condition evaluates as false. If we need more than one statement (i.e *statement_group*) then we need to use *begin-end* or a *fork-join* block. The *condition* here can be an expression or a single value. If the *condition* evaluates to '0' or unknown then the *condition* is considered false, and if the *condition* evaluates to '1' or more then the *condition* is considered true.

case	<u>Syntax</u>
	case(*expression*)
	    case_item1: *statement* or *statement_group*
	    case_item2: *statement* or *statement_group*
	
	    case_itemN: *statement* or *statement_group*
	    default: *statement* or *statement_group*
	endcase
	
	It compares the *expression* with each of the *case_item*'s and executes the *statement* or *statement_group* associated with the first matching *case_item*. It executes the *default* if none of the *case_item*'s match. Here the default case is optional. If we need more than one statement (i.e *statement_group*) then we need to use *begin-end* or a *fork-join* block.

Table 84: Programming Statements in Verilog

Programming Statements	Usage
casez	<ins>Syntax</ins> casez(*expression*) *case_item1*: *statement* or *statement_group* *case_item2*: *statement* or *statement_group* *case_itemN*: *statement* or *statement_group* *default*: *statement* or *statement_group* endcase It is special version of case statement where 'z' and '?' are treated as don't cares. Similar to case statement it compares the *expression* with each of the *case_item's* and executes the *statement* or *statement_group* associated with the first matching *case_item*. It executes the *default* if none of the *case_item's* match. Here the default case is optional. If we need more than one statement (i.e *statement_group*) then we need to use *begin-end* or a *fork-join* block.
casex	<ins>Syntax</ins> casex(*expression*) *case_item1*: *statement* or *statement_group* *case_item2*: *statement* or *statement_group* *case_itemN*: *statement* or *statement_group* *default*: *statement* or *statement_group* endcase It is special version of case statement where 'x', 'z' and '?' are treated as don't cares. Similar to case statement it compares the *expression* with each of the *case_item's* and executes the *statement* or *statement_group* associated with the first matching *case_item*. It executes the *default* if none of the *case_item's* match. Here the default case is optional. If we need more than one statement (i.e *statement_group*) then we need to use *begin-end* or a *fork-join* block.
forever	<ins>Syntax</ins> forever *statement* or *statement_group* It is an infinite loop that continuously executes the *statement* or *statement_group*. If we need more than one statement (i.e *statement_group*) then we need to use *begin-end* or a *fork-join* block.

Table 84: Programming Statements in Verilog

Programming Statements	Usage
repeat	Syntax repeat(*expression*) *statement* or *statement _group* Like forever it is a loop that executes the *statement* or *statement _group* a fixed number of times based on the expression. If we need more than one statement (i.e *statement_group*) then we need to use *begin-end* or a *fork-join* block.
while	Syntax while(condition) *statement* or *statement_group* It executes the statement or statement_group as long as the condition evaluates as true. If the *condition* evaluates to '0' or unknown then the *condition* is considered false, and if the *condition* evaluates to '1' or more then the *condition* is considered true. If we need more than one statement (i.e *statement_group*) then we need to use *begin-end* or a *fork-join* block.
for	Syntax for(*initial_value*; *condition*; *step*) *statement* or *statement_group* The for loop here uses three expressions separated by semicolons to control the loop. The first expression *(initial_value)* is executed once before entering the loop the first time. The second expression *(condition)* is evaluated to determine if the contents of the loop (i.e *statement* or *statement_group*) should be executed. If the loop condition expression is true, the loop is entered. The final expression *(step)* is evaluated at the end of the loop. If we need more than one statement (i.e *statement_group*) then we need to use *begin-end* or a *fork-join* block.
disable	Syntax disable *group_name*; It discontinues execution of a named group of statements.

9. What is Full_Case and Parallel_Case?

Full_Case

This particular directive is used to inform the synthesis tool that the 'case' statement is fully defined, and that the output assignments for all unused cases are don't cares. if all cases are in fact specified, synthesis tool can recognize this automatically. The functionality between 'pre' and 'post' synthesized designs may or may not remain the same when using this directive.

Parallel_Case

This particular directive is used to inform the synthesis tool that all the case items are mutually exclusive i.e more than one case item can never be true. Here when a design does have overlapping cases then the functionality between 'pre' and 'post' synthesis designs will be different.

10. What is Moore and a Mealy State Machine?

Moore State Machine

Here outputs are only a function of the present state. A Moore State Machine is shown in the figure below.

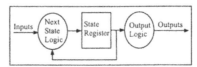

Figure 160: Moore State Machine

Mealy State Machine

Here one or more of the outputs are function of the present state and one or more of the inputs. A Mealy State Machine is shown in the figure below.

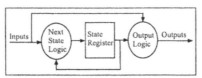

Figure 161: Mealy State Machine

11. What is a UDP?

A UDP (User Defined Primitives) describes a piece of logic with a truth table. They can be either combinatorial or sequential. They are scalar (1-bit) and only one output is allowed which must be the first terminal. The main reason UDP's are used is performance as Verilog evaluates UDP's quickly and UDP's take up a very small amount of memory. The most common use for UDP's is in modeling a library of standard cells. An optimistic 'mux' UDP is shown below. It is optimistic because if the inputs are same and the select is unknown, the input still propagates.

```
primitive mux(y, sel, a, b);
output y;
input a, b, sel;
   table
   // s  a  b  :  y
   0  0  ?  :  0;
   0  1  ?  :  1;
   1  ?  0  :  0;
   1  ?  1  :  1;
   x  0  0  :  0;
   x  1  1  :  1;
   endtable
endprimitive
```

12. What are Functions and Tasks in Verilog?

Verilog provides Functions and Tasks to allow the behavioral description of a module to be broken into more manageable parts allowing better readability and manageability. Functions and Tasks are useful for several reasons which are, they allow often-used behavioral sequences to be written once and called when needed, they allow for a cleaner writing style and finally they allow data to be hidden from other parts of the design.

Table below lists the characteristics of a Function and a Task.

Table 85: Function and Task

Function	Task
1. It returns a value to the expression that called it.	1. It does not return a value to an expression.
2. It takes zero time.	2. It can take more than zero time to execute.
3. It cannot contain delay or event controls (#, @ and wait).	3. It can contain delay or event controls (#, @ and wait).
4. It may be called from within a procedural and continuous assignment statements.	4. It cannot be called from a continuous assignment statement.
5. It has atleast one input but does not have outputs or inouts.	5. It can have zero or more arguments (i.e inputs, outputs or inouts) of any type.

6. Syntax for Function
 	
function [size_or_type] function_name;
   input declarations
   local variable declarations
   procedural_statement or statement_group
endfunction
	6. Syntax for Task
	
	task task_name;
	   input, output, and inout declarations
	   local variable declarations
	   procedural_statement or statement_group
	endtask

13. Provide Verilog code for the most commonly used Flops and Latches in a design?

Tables below provide the verilog definition for the most commonly used Flops and Latches in a design.

Table 86: Verilog Code for Regular/Synchronous/Asynchronous Flops

Regular Flop	Synchronous Reset Flop	Asynchronous Reset Flop

module FLOP(q, d, clk))
  input d, clk;
  output q;
  reg q;
    always @(posedge clk)
    q <= d;
  endmodule
	module SR_FLOP(q, d, reset, clk))
	  input d, reset, clk;
	  output q;
	  reg q;
	    always @(posedge clk)
	    if(!reset)
	      q <= 1'b0;
	    else
	      q <= d;
	  endmodule
		module AR_FLOP(q, d, reset, clk))
		  input d, reset, clk;
		  output q;
		  reg q;
		    always @(posedge clk or
		              negedge reset)
		    if(!reset)
		      q <= 1'b0;
		    else
		      q <= d;
		  endmodule

Table 87: Verilog Code for Regular/Synchronous/Asynchronous Latches

Regular Latch	Synchronous Reset Latch	Asynchronous Reset Latch

module LATCH(q, d, clk))
  input d, clk;
  output q;
  reg q;
    always @(clk or d)
    if(clk) q <= d;
  endmodule
	module SR_LATCH(q, d, reset, clk))
	  input d, reset, clk;
	  output q;
	  reg q;
	    always @(clk or d)
	    if(!reset) q <= 1'b0;
	    elseif(clk) q <= d;
	  endmodule
		module AR_LATCH(q, d, reset, clk))
		  input d, reset, clk;
		  output q;
		  reg q;
		    always @(clk or d or
		              negedge reset)
		    if(!reset) q <= 1'b0;
		    elseif(clk) q <= d;
		  endmodule

14. Provide different Verilog coding styles for a Mux functionality?

Table below provides the different Verilog coding styles for a Mux functionality.

Table 88: Verilog Coding Styles for a Mux Functionality

Mux Functionality	Using Continuous Assignment Statement	Using Procedural Block

1-bit, 2x1 Mux
 	
	Style1
	
	module MUX21(out, s, a, b)
	  input s, a, b;
	  output out;
	    assign out = s ? b : a;
	endmodule
	
	Style2
	
	module MUX21(out, s, a, b)
	  input s, a, b;
	  output out;
	    assign out = (s & b) | (!s & a);
	endmodule
		Style1
		
		module MUX21(out, s, a, b)
		  input s, a, b;
		  output out;
		  reg out;
		    always @(s or a or b)
		      if (s) out = b;
		      else out = a;
		endmodule
		
		Style2
		
		module MUX21(out, s, a, b)
		  input s, a, b;
		  output out;
		  reg out;
		    always @(s or a or b)
		      case(s)
		        1'b0 : out = a;
		        1'b1 : out = b;
		      endcase
		endmodule

4-bit, 4x1 one-hot Mux
 	
	Style1
	
	module MUX44(out, s, a, b, c, d)
	  input [3:0] s, a, b, c, d;
	  output [3:0] out;
	    assign out = s[0] ? a : (s[1] ? b : (s[2] ? c : d));
	endmodule
	
	Style2
	
	module MUX44(out, s, a, b, c, d)
	  input [3:0] s, a, b, c, d;
	  output [3:0] out;
	    assign out = ({4{s[0]}} & a) | ({4{s[1]}} & b)
	            ({4{s[2]}} & c) | ({4{s[3]}} & d);
	endmodule
		Style1
		
		module MUX44(out, s, a, b, c, d)
		  input [3:0] s, a, b, c, d;
		  output [3:0] out;
		  reg [3:0] out;
		    always @(s or a or b)
		      if (s[0]) out = a;
		      elseif(s[1]) out = b;
		      elseif(s[2]) out = c;
		      elseif(s[3]) out = d;
		endmodule
		
		Style2
		
		module MUX44(out, s, a, b, c, d)
		  input [3:0] s, a, b, c, d;
		  output [3:0] out;
		  reg [3:0] out;
		    always @(s or a or b)
		      case(s)
		        4'b0001 : out = a;
		        4'b0010 : out = b;
		        4'b0100 : out = c;
		        4'b1000 : out = d;
		        default : out = 4'bx;
		      endcase
		endmodule

Table 88: Verilog Coding Styles for a Mux Functionality

Mux Functionality	Using Continuous Assignment Statement	Using Procedural Block

4-bit, 4x1 encoded Mux 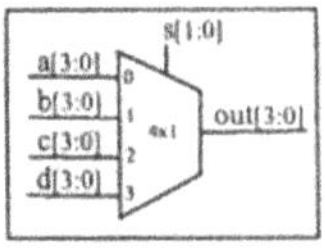	*Style1*
	
	module MUX44(out, s, a, b, c, d)
	  input [1:0] s;
	  input [3:0] a, b, c, d;
	  output [3:0] out;
	    assign out = (s = 2'b00) ? a : ((s = 2'b01) ? b :
	            ((s = 2'b10) ? c : d));
	endmodule
	
	Style2
	
	module MUX44(out, s, a, b, c, d)
	  input [1:0] s;
	  input [3:0] a, b, c, d;
	  output [3:0] out;
	    assign out = ({4{(!s[1] & !s[0])}} & a) |
	            ({4{(!s[1] & s[0])}} & b) |
	            ({4{(s[1] & !s[0])}} & c) |
	            ({4{(s[1] & s[0])}} & d);
	endmodule
		Style1
		
		module MUX44(out, s, a, b, c, d)
		  input [1:0] s;
		  input [3:0] a, b, c, d;
		  output [3:0] out;
		  reg [3:0] out;
		    always @(s or a or b)
		      if (s = 2'b00) out = a;
		      elseif(s = 2'b01) out = b;
		      elseif(s = 2'b10) out = c;
		      elseif(s = 2'b11) out = d;
		endmodule
		
		Style2
		
		module MUX44(out, s, a, b, c, d)
		  input [1:0] s;
		  input [3:0] a, b, c, d;
		  output [3:0] out;
		  reg [3:0] out;
		    always @(s or a or b)
		      case(s)
		        2'b00 : out = a;
		        2'b01 : out = b;
		        2'b10 : out = c;
		        2'b11 : out = d;
		      endcase
		endmodule

15. Write Verilog code for modelling different Memory configurations?

Table below shows Verilog code for different Memory configurations.

Table 89: Memory Models

Direct Mapped Cache(1 Read/Write Port) -

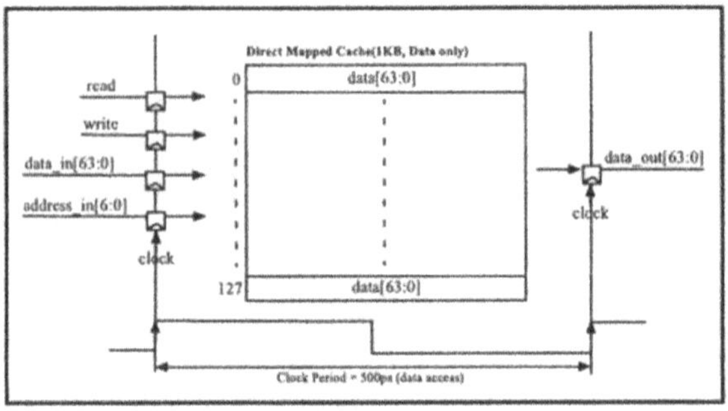

Figure 162: Direct Mapped Cache

Table 89: Memory Models

```
'timescale 1ps/1ps
module DIRECT_CACHE(read, write, data_in, address_in, data_out, clock)
 input read, write, clock;
 input [63:0] data_in;
 input [6:0] address_in;
 output [63:0] data_out;

 reg read_f, write_f, clock;
 reg [63:0] data_in_f;
 reg [6:0] address_in_f;
 reg [63:0] data_out, temp;
 reg [63:0] memory [0:127];

parameter period = 500;

//// Clock Generation ////
   initial
    begin
     clock = 0;
     forever #(period/2) clock = ~clock;
    end

//// Flop Incoming Control and Data Signals ////
   always @(posedge clock)
    begin
     read_f <= read;
     write_f<= write;
     data_in_f <= data_in;
     address_in_f <= address_in;
    end

//// Cache Write Operation ////
   always @(write_f)
    #1  if(write_f)
         memory[address_in_f] = data_in_f;

//// Cache Read Operation ////
   always @(read_f)
    #1
    if(read_f)
     temp = memory[address_in_f];
    else
     temp = 64'hffffffffffffffff;

//// Forward Data ////
   always @(posedge clock)
    data_out <= temp;
endmodule
```

Table 89: Memory Models

4-Way Set Associative Cache(1 Read/Write Port) -

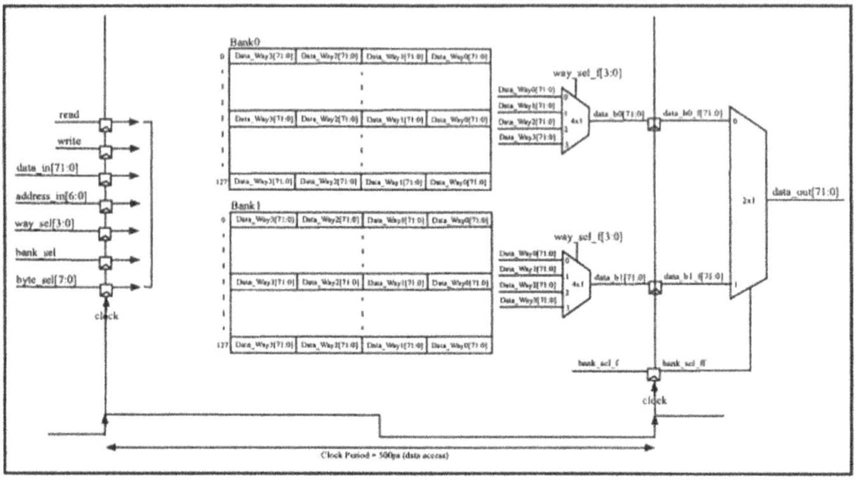

Figure 163: 4-Way Set Associative Cache

```
'timescale 1ps/1ps
module FWSA_CACHE(read, write, data_in, address_in, way_sel, bank_sel, byte_sel, data_out,
clock);
 input read, write, bank_sel, clock;
 input[71:0] data_in;
 input[6:0] address_in;
 input[3:0] way_sel;
 input[7:0] byte_sel;
 output[63:0] data_out;

 reg [71:0] array_b0_w0[0:127];
 reg [71:0] array_b0_w1[0:127];
 reg [71:0] array_b0_w2[0:127];
 reg [71:0] array_b0_w3[0:127];
 reg [71:0] array_b1_w0[0:127];
 reg [71:0] array_b1_w1[0:127];
 reg [71:0] array_b1_w2[0:127];
 reg [71:0] array_b1_w3[0:127];

 reg [71:0] temp0_b0_w0, temp0_b0_w1, temp0_b0_w2, temp0_b0_w3;
 reg [71:0] temp0_b1_w0, temp0_b1_w1, temp0_b1_w2, temp0_b1_w3;

 reg[71:0]temp0;
 reg [8:0] byte0, byte1, byte2, byte3, byte4, byte5, byte6, byte7;
 reg [71:0] data_b0, data_b1;
 reg [71:0] data_b0_f, data_b1_f;
 reg [71:0] data_out;
```

Table 89: Memory Models

```
  reg read_f, write_f;
  reg [71:0]data_in_f;
  reg [6:0] index_f;
  reg [3:0] way_sel_f, bank_sel_f, byte_sel_f;
  reg [7:0] bank_sel_ff;

parameter period = 500;

//// Clock Generation ////
  initial
   begin
    clock = 0;
    forever #(period/2) clock = ~clock;
   end

//// Flop Incoming Control and Data Signals ////
  always@(posedge clock)
   begin
    read_f <= read;
    write_f <= write;
    data_in_f <= data_in;
    index_f <= address_ in;
    way_sel_f <= way_sel;
    bank_sel_f <= bank_sel;
    byte_sel_f <= byte_sel;
    bank_sel_ff <= bank_sel_f;
   end

//// Cache Write Operation ////
  always @( write_ f or index_f or way_sel_f or bank_sel_f or byte_sel_f or data_in_f)
   #1
   begin
    if(write_f & ~bank_sel_f)      // Update Bank0
     begin
      if(way_sel_f == 4'b0001)      // Update Way0 of Bank0
       begin
        temp0 = array_b0_w0[index_f];
        byte0 = ({9{byte_sel_f[0]}} & data_in_f[8:0]) | ({9{~byte_sel_f[0]}} & temp0[8:0]);
        byte1 = ({9{byte_sel_f[1]}} & data_in_f[17:9]) | ({9{~byte_sel_f[1]}} & temp0[17:9]);
        byte2 = ({9{byte_sel_f[2]}} & data_in_f[26:18]) | ({9{~byte_sel_f[2]}} & temp0[26:18]);
        byte3 = ({9{byte_sel_f[3]}} & data_in_f[35:27]) | ({9{~byte_sel_ f[3]}} & temp0[35:27]);
        byte4 = ({9{byte_sel_f[4]}} & data_in_f[44:36]) | ({9{~byte_sel_f[4]}} & temp0[44:36]);
        byte5 = ({9{byte_sel_f[5]}} & data_in_f[53:45]) | ({9{~byte_sel_f[5]}} & temp0[53:45]);
        byte6 = ({9{byte_sel_f[6]}} & data_in_f[62:54]) | ({9{~byte_sel_f[6]}} & temp0[62:54]);
        byte7 = ({9{byte_sel_f[7]}} & data _in_f[71:63]) | ({9{~byte_se1_f[7]}} & temp0[71:63]);
        array_b0_w0[index_f] = {byte7, byte7, byte5, byte4, byte3, byte2, byte 1, byte0};
       end
      if(way_sel_f == 4'b0010)      // Update Way1 of Bank0
       begin
        temp0 = array_b0_w1[index_f];
```

Table 89: Memory Models

```
      byte0 = ({9{byte_sel_f[0]}} & data_in_f[8:0]) | ({9{~byte_sel_f[0]}} & temp0[8:0]);
      byte1 = ({9{byte_sel_f[1]}} & data_in_f[17:9]) | ({9{~byte_sel_f[l]}} & temp0[17:9]);
      byte2 = ({9{byte_sel_f[2]}} & data_in_f[26:18]) | ( {9 {~byte_sel_f[2]}} & temp0[26:18]);
      byte3 = ({9{byte_sel_f[3]}} & data_in_f[35:27]) | ({9{~byte_sel_f[3]}} & temp0[35:27]);
      byte4 = ({9{byte_sel_f[4]}} & data_in_f[44:36]) | ({9{~byte_sel_f[4]}} & temp0[44:36]);
      byte5 = ({9{byte_sel_f[5]}} & data_in_f[53:45]) | ({9{~byte_sel_f[5]}} & temp0[53:45]);
      byte6 = ({9{byte_sel_f[6]}} & data_in_f[62:54]) | ({9{~byte_sel_f[6]}} & temp0[62:54]);
      byte7 = ({9 {byte_sel_f[7]}} & data_in_f[71:63]) | ({9 {~byte_sel_f[7]}} & temp0[71:63]);
      array_b0_w1[index_f] = {byte7, byte7, byte5, byte4, byte3, byte2, byte1, byte0};
      end
    if(way_sel_f == 4'b0100)    // Update Way2 of Bank0
     begin
      temp0 = array_b0_w2[index_f];
      byte0 = ({9{byte_sel_f[0]}} & data_in_f[8:0]) | ({9{~byte_sel_f[0]}} & temp0[8:0]);
      byte1 = ({9{byte_sel_f[l]}} & data_in_f[17:9]) | ({9{~byte_sel_f[l]}} & temp0[17:9]);
      byte2 = ({9{byte_sel_f[2]}} & data_in_f[26:18]) | ({9{~byte_sel_f[2]}} & temp0[26:18]);
      byte3 = ({9{byte_sel_f[3]}} & data_in_f[35:27]) | ({9{~byte_sel_f[3]}} & temp0[35:27]);
      byte4 = ({9{byte_sel_f[4]}} & data_in_f[44:36]) | ({9{~byte_sel_f[4]}} & temp0[44:36]);
      byte5 = ({9{byte_sel_f[5]}} & data_in_f[53:45]) | ({9{~byte_sel_f[5]}} & temp0[53:45]);
      byte6 = ({9{byte_sel_f[6]}} & data_in_f[62:54]) | ({9{~byte_sel_f[6]}} & temp0[62:54]);
      byte7 = ({9{byte_sel_f[7]}} & data_in_f[71:63]) | ({9{~byte_sel_f[7]}} & temp0[71:63]);
      array_b0_w2[index_f] = {byte7, byte7, byte5, byte4, byte3, byte2, byte1, byte0};
      end
    if(way_sel_f == 4'b1000)    // Update Way3 of Bank0
     begin
      temp0 = array_b0_w3[index_f];
      byte0 = ({9{byte_sel_f[0]}} & data_in_f[8:0]) | ({9{~byte_sel_f[0]}} & temp0[8:0]);
      byte1 = ({9{byte_sel_f[1]}} &data_in_f[17:9]) | ({9{~byte_sel_f[1]}} & temp0[17:9]);
      byte2 = ({9{byte_sel_f[2]}} & data_in_f[26:18]) | ({9{~byte_sel_f[2]}} & temp0[26:l8]);
      byte3 = ({9{byte_sel_f[3]}} & data_in_f[35:27]) | ({9{~byte_sel_f[3]}} & temp0[35:27]);
      byte4 = ({9{byte_sel_f[4]}} & data_in_f[44:36]) | ({9{~byte_sel_f[4]}} & temp0[44:36]);
      byte5 = ({9{byte_sel_f[5]}} & data_in_f[53:45]) | ({9{~byte_sel_f[5]}} & temp0[53:45]);
      byte6 = ({9{byte_sel_f[6]}}& data_in_f[62:54]) | ({9{~byte_sel_f[6]}} & temp0[62:54]);
      byte7 = ({9{byte_sel_f[7]}} &data_in_f[71:63]) | ({9{~byte_sel_f[7]}} & temp0[71:63]);
      array_b0_w3[index_f] = {byte7, byte7, byte5, byte4, byte3, byte2, byte1, byte0};
      end

   if(write_f & bank_sel_f) // Update Bank1
    begin
     if(way_sel_f == 4'b0001)    // Update Way0 of Bank1
      begin
       temp0 = array _b1_w0[index_f];
       byte0 = ({9{byte_sel_f[0]}} & data_in_f[8:0]) | ({9{~byte_sel_f[0]}} & temp0[8:0]);
       byte1 = ({9{byte_sel_f[1]}} & data_in_f[17:9]) | ({9{~byte_sel_f[1]}} & temp0[17:9]);
       byte2 = ({9{byte_sel_f[2]}} & data_in_f[26:18]) | ({9{~byte_sel_f[2]}} & temp0[26:18]);
       byte3 = ({9{byte_sel_f[3]}} & data_in_f[35:27]) | ({9{~byte_sel_f[3]}} & temp0[35:27]);
       byte4 = ({9{byte_sel_f[4]}} & data_in_f[44:36]) | ({9{~byte_sel_fl4]}} & temp0[44:36]);
       byte5 = ({9{byte_sel_f[5]}} & data_in_f[53:45]) | ({9{~byte_sel_f[5]}} & temp0[53:45]);
       byte6 = ({9{byte_sel_f[6]}} & data_in_f[62:54]) | ({9{~byte_sel_f[6]}} & temp0[62:54]);
```

Table 89: Memory Models

```
        byte7 = ({9{byte_sel_f[7]}} & data_in_f[71:63]) | ({9{~byte_sel_f[7]}} & temp0[71:63]);
        array_b1_w0[index_f]= {byte7, byte7, byte5, byte4, byte3, byte2, byte1, byte0};
     end
   if(way_sel_f == 4'b0010)      // Update Way 1 of Bank1
     begin
      temp0 = array_b1_w1[index_f];
      byte0 = ({9{byte_sel_f[0]}} & data_in_f[8:0]) | ({9{~byte_sel_f[0]}} & temp0[8:0]);
      byte1 = ({9{byte_sel_f[1]}} & data_in_f[17:9]) | ({9{~byte_sel_f[1]}} & temp0[17:9]);
      byte2 = ({9{byte_sel_f[2]}} & data_in_f[26:18]) | ({9{~byte_sel_f[2]}} & temp0[26:18]);
      byte3 = ({9{byte_sel_f[3]}} & data_in_f[35:27]) | ({9{~byte_sel_f[3]}} & temp0[35:27]);
      byte4 = ({9{byte_sel_f[4]}} & data_in_f[44:36]) | ({9{~byte_sel_f[4]}} & temp0[44:36]);
      byte5 = ({9{byte_sel_f[5]}} & data_in_f[53:45]) | ({9{~byte_sel_f[5]}} & temp0[53:45]);
      byte6 = ({9{byte_sel_f[6]}} & data_in_f[62:54]) | ({9{~byte_sel_f[6]}} & temp0[62:54]);
      byte7 = ({9{byte_sel_f[7]}} & data_in_f[71:63]) | ({9{~byte_sel_f[7]}} & temp0[71:63]);
      array_b1_w1 [index_f] = {byte7, byte7, byte5, byte4, byte3, byte2, byte1, byte0};
     end
 if(way_sel_f == 4'b0100)       // Update Way2 of Bank1
   begin
    temp0 = array_b1_w2[index_f];
    byte0 = ({9{byte_sel_f[0]}} & data_in_f[8:0]) | ({9{~byte_sel_f[0]}} & temp0[8:0]);
    byte1 = ({9{byte_sel_f[1]}} & data_in_f[17:9]) | ({9{~byte_sel_f[1]}} & temp0[17:9]);
    byte2 = ({9{byte_sel_f[2]}} & data_in_f[26:18]) | ({9{~byte_sel_f[2]}} & temp0[26:18]);
    byte3 = ({9{byte_sel_f[3]}} & data_in_f[35:27]) | ({9{~byte_sel_f[3]}} & temp0[35:27]);
    byte4 = ({9{byte_sel_f[4]}} & data_in_f[44:36]) | ({9{~byte_sel_f[4]}} & temp0[44:36]);
    byte5 = ({9{byte_sel_f[5]}} & data_in_f[53:45]) | ({9{~byte_sel_f[5]}} & temp0[53:45]);
    byte6 = ({9{byte_sel_f[6]}} & data_in_f[62:54]) | ({9{~byte_sel_f[6]}} & temp0[62:54]);
    byte7 = ({9{byte_sel_f[7]}} & data_in_f[71:63]) | ({9{~byte_sel_f[7]}} & temp0[71:63]);
    array_b1_w2[index_f] = {byte7, byte7, byte5, byte4, byte3, byte2, byte1, byte0};
   end
  if(way_sel_f == 4'b1000)      // Update Way3 of Bank1
    begin
     temp0 = array _b1_w3[index_f);
     byte0 = ({9{byte_sel_f[0]}} & data_in_f[8:0]) | ({9{~byte_sel_f[0]}} & temp0[8:0]);
     byte1 = ({9{byte_sel_f[1]}} & data_in_f[17:9]) | ({9{~byte_sel_f[1]}} & temp0[17:9]);
     byte2 = ({9{byte_sel_f[2]}} & data_in_f[26:18]) | ({9{~byte_sel_f[2]}} & temp0[26:18]);
     byte3 = ({9{byte_sel_f[3]}} & data_in_f[35:27]) | ({9{~byte_sel_f[3]}} & temp0[35:27]);
     byte4 = ({9{byte_sel_f[4]}} & data_in_f[44:36]) | ({9{~byte_sel_f[4]}} & temp0[44:36]);
     byte5 = ({9{byte_sel_f[5]}} & data_in_f[53:45]) | ({9{~byte_se1_f[5]}} & temp0[53:45]);
     byte6 = ({9{byte_sel_f[6]}} & data_in_f[62:54]) | ({9{~byte_sel_f[6]}} & temp0[62:54]);
     byte7 = ({9{byte_sel_f[7]}} & data_in_f[71:63]) | ({9{~byte_sel_f[7]}} & temp0[71:63]);
     array_b1_w3[index_f] = {byte7, byte7, byte5, byte4, byte3, byte2, byte1, byte0};
    end

//// Cache Read Operation ////
  always @(read_f or indcx_f)
   #1
    if(read_f)
      begin
       temp0_b0_w0 = array_b0_w0[index_f];
       temp0_b0_w1 = array_b0_w1[index_f];
       temp0_b0_w2 = array_b0_w2[index_f];
```

Table 89: Memory Models

```
                  temp0_b0_w3 = array_b0_w3[index_f];
                  temp0_b1_w0 = array_b1_w0[index_f];
                  temp0_bl_wl = array_b1_w1[index_f];
                  temp0_b1_w2 = array_b1_w2[index_f];
                  temp0_bl_w3 = array_b1_w3[index_f];
                end
              else
                begin
                  temp0_b0_w0 = 72'hffffffffffffffffff;
                  temp0_b0_w1 = 72'hffffffffffffffffff;
                  temp0_b0_w2 = 72'hffffffffffffffffff;
                  temp0_b0_w3 = 72'hffffffffffffffffff;
                  temp0_b1_w0 = 72'hffffffffffffffffff;
                  temp0_b1_wl = 72'hffffffffffffffffff;
                  temp0_b1_w2 = 72'hffffffffffffffffff;
                  temp0_b1_w3 = 72'hffffffffffffffffff;
                end

//// Data Select for Bank0 ////
    always @(way_sel_f or temp0_b0_w0 or temp0_b0_w1 or temp0_b0_w2 or temp0_b0_w3)
    #2
    case(way_sel_f)
      4'b0001: data_b0 = temp0_b0_w0;
      4'b0010: data_b0 = temp0_b0_w1;
      4'b0100: data_b0 = temp0_b0_w2;
      4'b0100: data_b0 = temp0_b0_w3;
      default:data_b0=72'hxxxxxxxxxxxxxxxxxx;
    endcase
//// Data Select for Bank1 ////
    always @(way_sel_f or temp0_b1_w0 or temp0_b1_wl or temp0_b1_w2 or temp0_b1_w3)
    #2
    case(way_sel_f)
      4'b0001: data_b1 = temp0_b1_w0;
      4'b0010: data_b1 = temp0_b1_w1;
      4'b0100: data_b1 = temp0_b1_w2;
      4'bl000: data_b1 = temp0_bl_w3;
      default: data_b1 = 72'hxxxxxxxxxxxxxxxxxx;
    endcase

//// Flop 72-bit Data from Bank0 and Bank1 ////
    always @(posedge clock)
      begin
      data_b0_f <= data_b0;
      data_b1_f <= data_b1;
      end

//// Final Data Select ////
    always @(bank_sel_ff or data_b0_f or data_b1_f)
    #1
    data_out = bank_sel_ff ? data_b1 _f : data_b0_f;
endmodule
```

Table 89: Memory Models

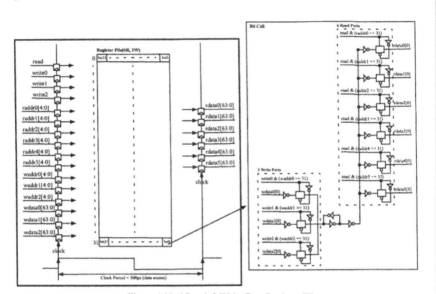

Figure 164: 6 Read, 3 Write Port Register File

```
'timescale 1ps/1ps
module 6R_3W_RF(read, write0, write1, write2, raddr0, raddr1, raddr2, raddr3, raddr4, raddr5,
waddr0, waddr1, waddr2, rdata0, rdata1, rdata2, rdata3, rdata4, rdata5, wdata0, wdata1, wdata2,
clock);
  input read, write0, write1, write2, clock;
  input [4:0] raddr0, raddr1, raddr2, raddr3, raddr4, raddr5;
  input [4:0] waddr0, waddr1, waddr2;
  input [63:0] wdata0, wdata1, wdata2;
  output[63:0] rdata0, rdata1, rdata2, rdata3, rdata4, rdata5;

  reg read_f, write0_f, write 1_f, write2_f;
  reg [4:0] raddr0_f, raddr1_f, raddr2_f, raddr3_f, raddr4_f, raddr5_f;
  reg [4:0] waddr0_f, waddr0_f, waddr0_f;
  reg [63:0] wdata0_f, wdata1_f, wdata2_f;
  reg [63:0] rdata0, rdata1, rdata2, rdata3, rdata4, rdata5;

  reg [63:0] mem[0:31];

parameter period = 500;
```

Table 89: Memory Models

```
//// Clock Generation ////
  initial
    begin
      clock = 0;
      forever #(period/2) clock = ~clock;
    end

//// Flop Incoming Control and Data Signals ////
  always@(posedge clock)
    begin
      read_f <= read;
      write0_f <= write0;
      write1_ f <= write1;
      write2_f <= write2;
      raddr0_f <= raddr0;
      raddr1_f <= raddr1;
      raddr2_f <= raddr2;
      raddr3_f <= raddr3;
      raddr4_f <= raddr4;
      raddr5_f <= raddr5;
      waddr0_f <= waddr0;
      waddr1_f <= waddr1;
      waddr2_f <= waddr2;
      wdata0_f <= wdata0;
      wdata1_f <= wdata1;
      wdata2_f <= wdata2;
    end

//// Write Operation ////
  always @(write0_f or write1_ f or write2_f or waddr0_ f or waddr1_f or waddr2_f or wdata0_f or
wdata1_f or wdata2_f)
    #1
    begin
      if(write0_f) mem[waddr0_f] = wdata0_f;
      if(write1_f) mem[waddr1_f] = wdata1_f;
      if(write2_f) mem[waddr2_f] = wdata2_f;
    end

//// Read Operation ////
  always @(negedge clock)
    if(read_f)
      data0 = mem[raddr0_f];
      data 1 = mem[raddr1_ f];
      data2 = mem[raddr2_f];
      data3 = mem[raddr3_f];
      data4 = mem[raddr4_f];
      data5 = mem[raddr5_f];
```

Table 89: Memory Models

```
    else
      data0 = 64'hffffffffffffffff;
      data1 = 64'hffffffffffffffff;
      data2 = 64'hffffffffffffffff;
      data3 = 64'hffffffffffffffff;
      data4 = 64'hffffffffffffffff;
      data5 = 64'hffffffffffffffff;

//// Forward Read Data ////
    always @(posedge clock)
      begin
       rdata0 <= data0;
       rdata1 <= data1;
       rdata2 <= data2;
       rdata3 <= data3;
       rdata4 <= data4;
       rdata5 <= data5;
      end
endmodule
```